Leonardo Mendoza

Design of Pentane Refrigeration System with Propane Refrigerant

Leonardo Mendoza

Design of Pentane Refrigeration System with Propane Refrigerant

Design of a Tube Housing Heat Exchanger

ScienciaScripts

Imprint

Any brand names and product names mentioned in this book are subject to trademark, brand or patent protection and are trademarks or registered trademarks of their respective holders. The use of brand names, product names, common names, trade names, product descriptions etc. even without a particular marking in this work is in no way to be construed to mean that such names may be regarded as unrestricted in respect of trademark and brand protection legislation and could thus be used by anyone.

Cover image: www.ingimage.com

This book is a translation from the original published under ISBN 978-620-2-15997-5.

Publisher:
Sciencia Scripts
is a trademark of
Dodo Books Indian Ocean Ltd. and OmniScriptum S.R.L publishing group

120 High Road, East Finchley, London, N2 9ED, United Kingdom
Str. Armeneasca 28/1, office 1, Chisinau MD-2012, Republic of Moldova, Europe
Printed at: see last page
ISBN: 978-620-6-42426-0

Table of Contents :

DESIGN OF THE PENTANE COOLING SYSTEM OF THE LPG FRACTIONATION PLANT BAJO GRANDE

AUTHOR

Br. Leonardo J. Mendoza S.

Br. Merlin J. Barreto O.

SUMMARY

The purpose of the research was to design a refrigeration system for the large fractionation plant that would allow adjusting the temperature conditions of the pentane to be stored in the plant. For the execution of the design it was necessary to describe the plant's refrigeration system in order to select the most suitable location for the new pentane cooler. The second phase consisted of establishing the operational and design parameters under which the new chiller will operate, for which it was necessary to consult engineering manuals, personnel from the Bajo Grande plant, La Salina refrigeration plant and design standards. Subsequently, calculations were started to determine the dimensions of the refrigeration system. This research is considered to be of the non-experimental projective documentary and field type. The data collection techniques were direct observations, unstructured interviews with plant personnel and literature review. The results obtained allowed the optimization of the pentane production in the large fractionation plant, since the proposal is considered feasible.

Keywords: Refrigeration System, Chiller, Pentane, Heat Transfer, Heat Transfer

INTRODUCTION

Refrigeration systems are very common in industries involved in natural gas processing and processes related to petroleum refining, petrochemical and chemical industries. Among these uses are the use of refrigeration systems for NGL and LPG recovery as well as for hydrocarbon dew point control, reflux condensation in fractionation towers and liquefied natural gas (LNG) plants.

The operation of a typical cooling system of a natural gas processing plant is generally determined by the use of heat exchangers, which are equipment composed of a casing or body and an internal arrangement of tubes or pipe segments to achieve thermal exchange between fluids.

The cooling fluid is passed through the tube side and the product to be cooled through the casing side, the result is the temperature exchange between fluids, thus achieving the desired temperature. The selection of the cooling fluid depends on temperature requirements, availability, economy and previous experience.

The refrigeration system at the Bajo Grande LPG Fractionation Plant is a closed system that uses propane as the refrigerant and a three-stage centrifugal compressor (K-301A/B/C/D). The function of this system is to cool the fractionated products in the plant prior to their storage in refrigerated tanks.

Pentane and its isomers are components of gasoline and various volatile fractions from petroleum distillation. They are used in numerous synthesis processes, including the manufacture of intermediates, carrier agents and propellants, in polymerization reactions and as extraction solvents, as well as in the manufacture of low-temperature thermometers.

In Venezuela, specifically in the western region, pentane is contained in the natural gasoline produced in the main plants of the region, therefore its use depends on the start of its production.

The CCO will send a flow of pentane to the large fractionation plant for cooling. In this sense, it can be indicated that an infrastructure for cooling pentane is required since the plant does not currently have a cooling system for this product.

Pentane should be stored at temperatures below ambient temperature, since according to the GPSA Pressure-Enthalpy diagram it has a bubble point at atmospheric pressure of approximately 97 °F (36 °C), which means that under ambient conditions the product evaporates. For this reason it must be refrigerated to avoid product losses to the atmosphere and to guarantee the stability of the process.

Given the lack of a pentane refrigeration system at the Bajo Grande fractionation plant, and taking into account the need mentioned above, it is considered imperative to design such a system that allows adjusting the temperature to that required at the plant.

The following research work consists of 4 chapters. The first chapter deals with the problem statement; it describes in a broad manner the problem to be studied, its objectives, justification and importance of our project and the delimitation that may influence the realization of this research.

Chapter 2 contains the theoretical framework, which contains in a very complete way the previous background to this project, all the theoretical bases necessary to understand the study in question and some definitions to facilitate the understanding of the same.

Chapter 3 is the Methodological Framework, which defines the type of research, the research design, as well as the sources and techniques for data collection, and the unit of analysis.

CHAPTER I

THE PROBLEM

Problem statement

Natural gas is a valuable energy source that can be used to obtain derivatives, as fuel or to produce energy. It also represents an alternative to replace vehicle fuel under a cleaner fuel scheme. Currently in the world there are proven reserves that exceed 200 trillion cubic feet, which opens the doors of the economy of the nations that possess this valuable energy source towards a better future.

Natural gas is formed by the most volatile members of the paraffinic series of hydrocarbons, mainly methane, minor amounts of ethane, propane, butane and finally it may contain very small percentages of heavier compounds (Martinez 1994).

The proven gas reserves in our country are characterized by being associated with oil, which is why a large part of the gas production is injected or vented. It is worth noting as clear evidence of Venezuela's late interest in natural gas, that it was not until 1969 that a decree regulating the conservation of hydrocarbon resources was approved, whose main objective was to regulate the burning of gas in the so-called burners (PDVSA, 2005). (PDVSA, 2005).

The importance of natural gas lies in obtaining natural gas liquids (NGL), which are made up of propane, butane, pentane and other heavier hydrocarbon components, and are used in the domestic market as fuel and raw material, as well as in the domestic market. On the other hand, 31% of production supplies international markets. Venezuela's most important industry, PDVSA, through its subsidiary PDVSA Gas, is focused on extracting, fractioning, storing, shipping and marketing natural gas liquids (NGL). Thus, when the gas is processed, the resulting compounds (NGL) are fractionated, that is, the mixture is separated into individual products, which must be refrigerated within specifications for storage and subsequent shipment to vessels (PDVSA Gas 2010).

The Bajo Grande LPG Fractionation Plant, which belongs to PDVSA Gas and is part of the Western Gas Processing Department, is responsible for the fractionation of Natural Gas Liquids (NGL) and their refrigeration, through a system that uses propane as a refrigerant and consists of three refrigeration levels: 32 °C, 0 °C and -26 °C. The propane is stored in the accumulator drum V-311 and from there is fed to the cooler casings E-305 (propane), E-318 (n-butane), E-316 (iso-butane) and E-307 (gasoline).

This refrigeration system consists of: A propane refrigeration system, an iso-butane one, a normal-butane refrigeration system and finally a natural gasoline one, each of these systems are coupled to a supply line of a refrigerant (propane) coming from an accumulator tank, this system operates under a closed cycle scheme.

It is worth mentioning that the Bajo Grande Fractionation Plant was built between 1968 and 1969 and started operating in June 1970. It is located in the municipality of La Cañada de Urdaneta, south of the city of Maracaibo on an area of 195 hectares and collects natural gas liquids (L.P.G.) from the Lama and Lamarlíquido plants; and butane and heavier mixtures supplied by the El Tablazo I and II NGL plants.

This feed is processed to obtain propane, iso-butane, n-butane and natural gasoline which comprises the pentane fraction (**C5+**). These products serve as feedstock for the petrochemical and refining industries. Functionally, the Bajo Grande LPG plant consists of three operational sections: Fractionation, Storage and Product Dispatch. Originally, this plant was designed to produce 11505

BPD of propane, 5173 BPD of iso-butane, 9526 BPD of n-butane, and 5538 BPD of natural gasoline, since in addition to the feedstock from the Lama, Lamarlíquido, and El Tablazo plants, there was a butane feed from the La Paz plant. Subsequently, when the La Paz plant went out of service in 1987, the Bajo Grande LPG plant had the capacity to process 25600 BPD of LPG to obtain 11505 BPD of **C3**, 2567 BPD of **i-C4**, 5990 BPD of n-C4 and 5538 BPD of C5+.

The final product obtained in the fractionation process, natural gasoline, is composed of pentane (Iso and Normal pentane) and other heavier hydrocarbons, which means that the gasoline produced in the plant has a high RVP of 17 psia (Unstabilized Natural Gasoline) due to the pentane content.

In this regard, it was found that the problem lies in the need to reduce the value of the Reid Vapor Pressure (RVP) of gasoline to a maximum value of 7 Psia, in order to comply with international requirements for regulating the emission of organic vapors into the atmosphere, and this is achieved by removing the pentane content (PDVSA Gas 2010).

Pentane is a particular product in the production of liquids from the NGL extraction industry, as it is commonly contained in natural gasoline (C5+). When these are produced they can be technically treated as N-butane. With the above, it can be indicated that pentane can be stored as a refrigerated product at atmospheric pressure or as a pressurized product.

As a refrigerated product, it can be stored in fixed roof (or double floating roof) tanks at pressures around 0.1 Psig and temperatures of approximately 75°F, for which a refrigeration system is required.

As a pressurized product, pentane in large volumes, at pressures of 25 Psig and room temperature, must be stored in spheres. Since the technical-economic feasibility points to the option of refrigerated storage of the pentane that will be produced at the plant, and the flow coming from the New Fractionation Train of the Western Cryogenic Complex (NTF-CCO), a system that allows its cooling is required.

As a result, it was found that the problem lies in the fact that the large low fractionation plant does not have a pentane cooling system, due to this situation arises the need to design a pentane cooling system for the LPG Fractionation Plant Bajo Grande, which can be coupled to the propane refrigerant supply line, and put the pentane in required specifications, since the plant does not have a pentane cooler in its refrigeration system.

The pentane product is visualized to be used as: Feed for the Petrochemical Industry Paraguana Refinery, Feed for the Petrochemical Industry Olefins II Conversion, Fuel for Turbine Generators, Export Market, etc. At present, the existing plants in the western region do not have facilities for the storage of pentane, which makes imperative the need for the design and construction of new facilities for the storage and dispatch of this product. It should be noted that it was previously directed to the La Salina Lake Terminal located in Cabimas, Zulia state, for refrigeration and storage.

However, studies later determined that it was not strategically and economically viable to carry out this process in La Salina, since pentane will be produced in the Bajo Grande fractionation plant and it is more feasible to refrigerate it in the plant itself, since there is physical space to place new equipment and/or containers.

Problem Formulation

Within this framework, and based on the situation presented, it is convenient to ask the following question:

What are the characteristics that a pentane refrigeration system must have in the Bajo Grande fractionation plant, which will allow to adapt the pentane temperature conditions for its storage?

Research Objectives

General Objective

Design a pentane refrigeration system at the Bajo Grande LPG fractionation plant.

Specific Objectives

• Describe the Iso-butane, Normal-butane, and Natural Gasoline refrigeration systems existing in the LPG fractionation plant under large.

• Establish the operational and design parameters of the new pentane refrigeration system at the Bajo Grande LPG plant.

• Plantear los elementos (chiller y líneas asociadas) del nuevo sistema de refrigeración de pentano de la planta de fraccionamiento GLP bajo grande.

• Propose a pentane refrigeration system for the LPG fractionation plant under large.

Research Justification

As a result of the obsolescence of pentane production plants in the western region, and therefore of pentane cooling systems, there is an urgent need to create a system that allows its cooling, in this sense the search for theories and concepts, among other aspects, will be carried out to support the design of such a system. For this reason, this work serves as a basis for future research that can be carried out in the institution, given the variety of existing information.

Likewise, this instrument serves as a guide for future researchers who have the need to create a similar system and do not have the required tools at hand to start, and even more so if they have a vast idea about it. It also allows to add one more instrument to continue giving bases to gas engineering that little by little is receiving necessary contributions that cover the scarcity of information, given the relatively new creation of the career in the country.

On the other hand, it has been possible to investigate that much of the information concerning the gas engineering career, is defined in other languages, since foreign countries have an extensive trajectory started a long time ago, which has allowed writing field experiences and research that serve as a tool for the implementation of procedures, equipment design, design of refrigeration systems, natural gas processing, among others.

From an ecological point of view, it is also important to prevent the release of unstabilized gasoline vapors into the atmosphere, which causes atmospheric pollution by venting these vapors due to the increased pressure in the storage tank.

On the other hand, from the technical point of view, the importance of the design of the refrigeration system is highlighted, since the properties of pentane allow its evaporation under normal (ambient) conditions and consequently, when vapors are present inside the storage tank, this causes the pressure to rise, which in turn leads to the activation of the relief valves, calibrated to withstand a certain pressure, to vent the excess vapor into the atmosphere.

It should be noted that for the storage of pentane, two alternatives were studied, under which it is possible to store pentane and which determined the selection of the best option: A) refrigerated storage at atmospheric pressure or B) as a pressurized product.

For option "A", its economic and operational feasibility was determined, based on the requirements to be incorporated for the storage of the product, for which two tanks of 90000 bbl C/U of fixed roof with insulation are required and has an investment of 29.94 MMBsF.

Option B, i.e., pressurized product at ambient temperature, requires six (6) 30000 bbl sphere tanks, which has an investment of 64.19 MMBsF. Consequently, the best alternative is refrigerated storage (option A) since on the one hand there is a lower infrastructure cost, and on the other hand it is highlighted that this project does not generate income since a storage system is an operational support infrastructure, therefore it will be a cost only project and the least expensive alternative will be the most favorable option. Consequently, the chosen option represents a compelling reason that justifies the present study.

Also, the selection of the refrigerated storage of pentane is due to the fact that studies carried out by PDVSA Gas personnel revealed the feasibility of this system, which in turn seeks the adequate processing of pentane, within the framework of projects for hydrocarbon production that PDVSA Gas has in order to satisfy market demands, as is the case of the "Complejo Criogénico de Occidente" project, which will send approximately 7800 BPD of pentane to the Bajo Grande plant for its refrigeration, in addition to 1400 BPD that are estimated to be produced in the plant within the framework of the installation of the new depentanizing unit. For this reason, it is necessary to design a pentane cooling system.

Research Delimitation

Space

This work will be carried out at the Bajo Grande Fractionation Plant, which is part of the Western Gas Processing Management and belongs to PDVSA Gas, and is located on the side of the road via La Cañada, at km 16, in the Municipality of La Cañada de Urdaneta, Zulia State. It is in charge of the Fractionation of Natural Gas Liquids (NGL), for the production of Propane, Isobutane, Normalbutane and Natural Gasoline.

Temporary

The estimated time for the execution and development of the research work is six (6) months, beginning on May 4, 2011 and ending on November 11, 2011.

CHAPTER II

THEORETICAL FRAMEWORK

Research background

Hernández (2011), realizó el "Diseño de un sistema de refrigeración aplicado a la descarga de propano de buques en la Planta de Refrigeración La Salina". The general objective of the research was to design a refrigeration system applied to the unloading of vessels, to optimize gas production at the plant, avoiding gas flaring in the mechurrio.

The research was of the projective type, since solutions were proposed to solve the problem, it was also considered descriptive. The research design was non-experimental, since the refrigeration system is based on the procedures established in the plant.

As a result, the optimization of the ship unloading process was obtained, minimizing propane losses during the cabotage process and reducing environmental pollution, and therefore it is considered feasible.

This research added important elements to the current investigation, since useful procedures and methods are obtained to design the pentane refrigeration system for the fractionation plant under large.

Castillo and Nava (2005), carried out the "Design and construction of a cooling system for the liquefaction of toxic gases contained in pressurized containers".

The purpose of the study was to perform a complete design of the cooling system, building a cooling tank containing calcium chloride and where the cylinder, the evaporator, and selected other components such as the condensing unit, the expander device, the insulator to be used will be immersed.

The agitation system where the field tests are performed is where an evaporator coil is used to cool the calcium chloride, using Freon 22 as the refrigerant in the system.

The system built meets the need to cool the calcium chloride to -20° C in less than 6 hours, bringing the chloride to the temperature required for liquefaction, thus minimizing possible risks at the moment of valve extraction.

Consequently, this research is of an experimental, field type. It provides technical data about the design of the cooling system.

Torrez (2011), conducted the "Evaluation of the feasibility of using exchangers of the LPG-1 train fractionation area as part of the LPG-2 train cooling system". This study was focused on the feasibility of using the LPG-2 train cooling system due to the failure of the LPG-1 train exchangers at the ULÉ fractionation plant.

The general objective of the research was to obtain results about the behavior of the heat exchangers when subjected to operational conditions different from those in which they were designed, all with the purpose of optimizing propane production in the plant.

The proposal was to use the E10-A/B exchangers of the LPG-1 train as a replacement for the D2-509/510 (out of service) of the LPG-2 train. The results obtained were positive, the E10-A/B exchangers presented acceptable operational performance since they can cool propane and butane to the required temperatures.

This study provides a clear idea about the evaluation of a proposal, operational and economic feasibility of a system, looking for the best options to carry out a process.

This research is considered to be of the projective type since it attempts to propose a solution to a situation based on a previous research process.

PDVSA GAS Reference Frame

Historical Review

1953: The Ministry of Mines and Hydrocarbons creates the National Petrochemical Directorate to contribute to promote economic development through the industrialization of natural gas.

1970: The Anaco - Puerto Ordaz gas pipeline is completed, with a length of 228 km. This pipeline will supply natural gas to the Orinoco Mining Co. Siderúrgica de Orinoco and other industries installed in the industrial zone of Puerto Ordaz. Construction of the central gas pipeline of the lake begins.

1974: the Gas Processing Plant (LPG) with a processing capacity of 165 MMPCED was inaugurated at the Tablazo petrochemical complex in the state of Zulia.

1975: part of the Morón - Barquisimeto gas pipeline was built, which constitutes the first expansion of this important gas pipeline. On August 29, 1975, in an act held only in the elliptical hall of the federal capitol, the President of the Republic of that time, gave the "Cúmplase" to the organic law that reserves to the state the industry and commerce of hydrocarbons.

1982: Construction work began on the Oriente cryogenic complex, which consists of an extraction plant in San Joaquín, a 96 km (16") polyduct and a fractionation plant in Jose, Anzoátegui State. It also has an 8" and 56 Km. polyduct between Jose and the Puerto la Cruz refinery.

1983: The construction of the Quiriquire - Maturín gas pipeline is completed, with a length of 49.1 Km (20" diameter) and a capacity of 200 MMPCD. It will supply gas to the industrial zone of Maturín, La Toscaza and Jusepín.

1985: the Oriente Cryogenic Complex begins operations. The San Joaquín extraction plant has a capacity of 23 MMPCD of gas and the fractionation plant has a capacity of 70,000 BPD.

1986: The Gas Management was restructured and incorporated into its functions the operation of the Oriente cryogenic complex and the transfer to Lagoven and Maraven of the gas to sales activities carried out to date by Corcoven in the western part of the country.

1992: at the San Joaquín extraction plant and Jose fractionation plant, processing capacity was increased from 800 to 1000 MMPCED and from 70 to 100 thousand barrels per day, respectively. During the year, 10 measuring stations were added to the national gas pipeline network in order to optimize gas transmission and distribution processes.

1993: The first phase of the expansion of the Oriente cryogenic complex (ACCRO) was completed, increasing processing capacity to 1000 MMPCD and 100,000 BPD of NGLs, with external financing of 442 MM$ and a total investment of 46,758 MMBs.

1997: In the second half of 1997, PDVSA began a transformation process aimed at creating value and undertaking an organizational restructuring with an impact on the management of business processes.

1998: On January 1, PDVSA Gas, a subsidiary of Petróleos de Venezuela, began operations as part of the manufacturing and marketing division. Its responsibility is to promote the natural gas business in the country, for which it develops processing, transportation and distribution activities with other companies for the placement and sale of natural gas, achieving the harmonious integration of cultures and work teams in the East and West.

1999: the organic law for gaseous hydrocarbons was enacted, which defines the legal framework required to support the business throughout the value chain. MEM approval was also obtained for the Anaco area fields in the gas district of Venezuela.

2000: continued with the incorporation of the legal framework of the gas industry through the approval on May 31 of the regulation of the organic law of gaseous hydrocarbons (RLOHG) and the preparation, together with the MEM, of the proposal for the organization of the gas regulator.

2001: During the year, PDVSA Gas, S.A. was consolidated as a vertically integrated company. As a vertically integrated company, having completed the transfer of personnel, operating fields in the Anaco district and Block E South of the lake. The business portfolio was defined in line with the national gas plan and relations with PDVSA's subsidiaries were strengthened for the execution of service agreements.

2002: the conflict situation generated as of December 2 by a numerous group of oil industry workers, generated the closing of oil wells and consequently the production of associated gas, this situation contributed to restrict the supply of natural gas or methane from our network systems and of liquefied petroleum gas (LPG) or propane to the filling plants of cylinders for the residential and commercial/industrial sector that use this type of containers.

These events caused significant losses for PDVSA Gas and therefore for Venezuela. The so-called "oil strike" affected the major areas of PDVSA Gas, such as the operational, commercial and administrative areas. In the first area there was a significant decrease in the availability of methane gas and domestic and commercial LPG, due to the low contribution of gas associated with the shutdown of oil production as a result of high crude inventories, as well as operational problems in the San Joaquín extraction plant, which reduced its capacity by 50%, with the consequent decrease in the volumes of gas and NGL processed, this situation, in turn, limited methane gas deliveries to the domestic market.

These events triggered the non-compliance with contractual commitments with customers, due to the limited availability of gas, it was distributed on a priority basis to the domestic and electricity sectors and basic companies in Guayana.

In the administrative area, telecommunications services, servers, mail, on-site support, SAP and other financial systems were suspended, and billing and collection processes were paralyzed, which prevented the recovery of the value of methane gas and NGL sales.

In the second week of December, "El Palito, Bajo Grande and Ulé" were affected by the shutdown of the Paraguaná refining center and the El Palito refinery, affecting the supply of LPG in cylinders for the communities and collectives in general. The third week was the most critical period and the dispatches of LPG cylinders were made from Jose for the whole country.

In the fourth week of December, supply slowly began to be reestablished with the incorporation of Guatire and El Guamache (Margarita Island). As a consequence of this situation, gas production dropped to 856 MMPCD; methane gas deliveries to the sales system stood at 1065 MMPCD; while NGL production and sales dropped to 43 and 57 thousand barrels per day, respectively.

2003: In January, 5 LPG supply sources were stabilized: Jose, Guatire, Bajo Grande, Puerto la Cruz and El Guamache. However, the supply of Ulé, Cardón and El Palito was not normalized. In order to achieve this objective, a series of actions were defined, among which the following stand out: implementation of a transitory organization that would allow the company to undertake the most priority tasks and establish a sense of direction for the personnel; incorporation of temporary personnel to reinforce the weakest areas and establishment of work and contingency plans to give continuity to the core activities of the business, such as: operational control, billing and collections, finance, telecommunications and information technology and public affairs. As a result of these actions and the great effort made by all personnel, PDVSA Gas has been able to recover its operating levels to date.

Due to the oil strike in December 2002, the payment process of the industry's large customers (failures in the collection and invoicing system), especially in the steel, iron and steel and petrochemical sectors, created congestion and problems in invoicing.

PDVSA Gas was operating with 98% of its processes. Domestic consumption was fully supplied, receiving 1700 MMPCDM from the production sector, completing 100% of the domestic market demand.

Record gas sales were recorded in the systems managed by Anaco-Barquisimeto, Anaco-Puerto Ordaz and Anaco-Puerto la Cruz (Jose), with a value of 1696 MMPCD, in the month of April, surpassing the previous record of 1690, established in December 2001.

2004: the new PDVSA Gas signed the bid for phase 1 of the ICO (Interconexión Centro Occidente) project for the start of construction of the Quero - Río Seco section.

In April, the record for methane gas sales in the domestic market was once again surpassed, with an increase of 1979 MMPCD compared to March of this year. PDVSA Gas began construction of the early phase of the West Central interconnection project (ICO), which consists of a 70 km pipeline of 36" diameter pipeline.

2005: By resolution of the Board of Directors of Petróleos de Venezuela S.A., it was agreed to integrate the Gas business at the National level, based on an action plan.

Theoretical bases of the research

Pentane

It is a hydrocarbon or alkane with chemical formula **C5H12.** The word pentane can also refer to its structural isomers, or a mixture of them. In IUPAC nomenclature, however, pentane is the exclusive name for *n-pentane*; the other two (2) chain isomers are named methylbutane and dimethylpropane.

Most frequent uses of pentane

Pentane and its isomers are components of gasoline and various volatile fractions from petroleum distillation. They are used in numerous synthesis processes, including the manufacture of intermediates, in polymerization reactions and as extraction solvents.

Its most important application is in the gasoline production industry. In the case of the oil industry PDVSA Gas, the pentane product is visualized to be used as feed for the petrochemical industry Paraguaná Refinery, feed for the Petrochemical Industry Conversion Olefins II, Fuel for Turbine Generators, Export Market and for the recovery of heavy crude oil.

Physical and thermodynamic properties of pentane

Some physical and thermodynamic properties of pentane can be determined from the pressure-enthalpy diagram described below:

Pressure-Enthalpy Diagram of Pentane

The pressure-enthalpy diagram is a graph used to determine the volumes, phases, entropies, enthalpies, specific volumes of a hydrocarbon or non-hydrocarbon gas (Fig. 1). The left part of the graph is the saturated liquid region, and its limit is up to the center of the curve (Dome) that divides the phases. The right region corresponds to the superheated vapor phase, while the area under the curve represents the mixing region. GPSA (2004)

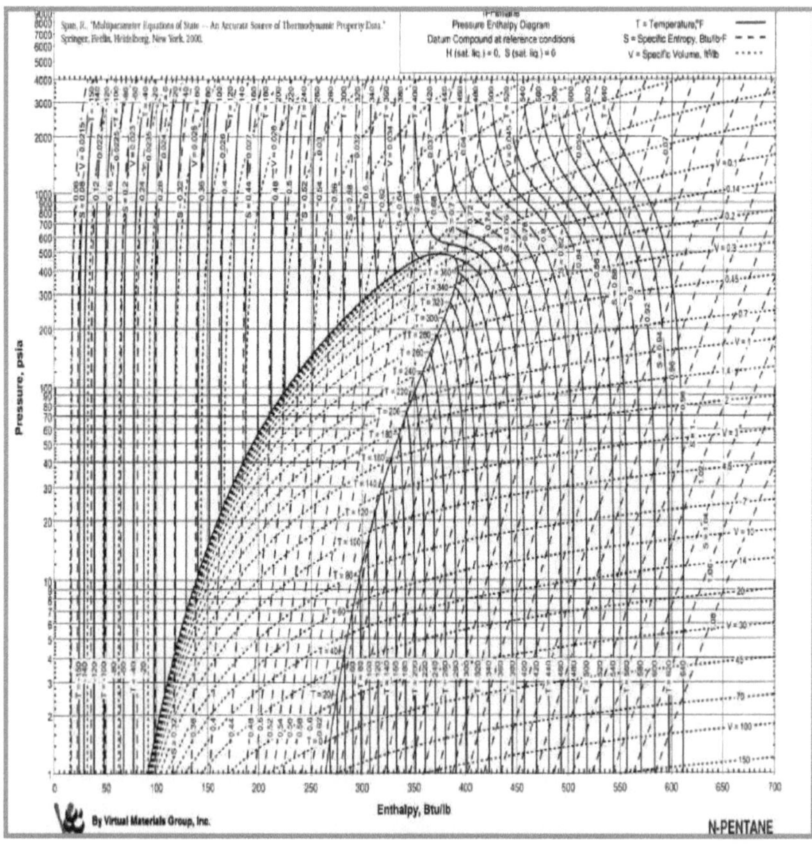

Figure 1. **Pressure-Enthalpy Diagram of Pentane.** Source GPSA 2004.

Saturation pressure of pentane.

Saturation pressure is the pressure at which a substance begins to boil at a given temperature. Pentane begins to evaporate at 97 °F (36 °C) at a saturation pressure of 14.7 psia. Cengel (2009)

Saturation temperature of pentane:

The temperature at which a substance begins to evaporate at a given pressure. Pentane at 14.7 psia has a saturation temperature of 97°F. Cengel (2009)

Specific heat of pentane (Cp)

Specific heat is defined as the energy required to raise the temperature of a unit mass of a substance by one degree. In the case of pentane the Cp at ambient conditions is 0.5439 (BTU/Lb °F) in the liquid state. GPSA (2004)

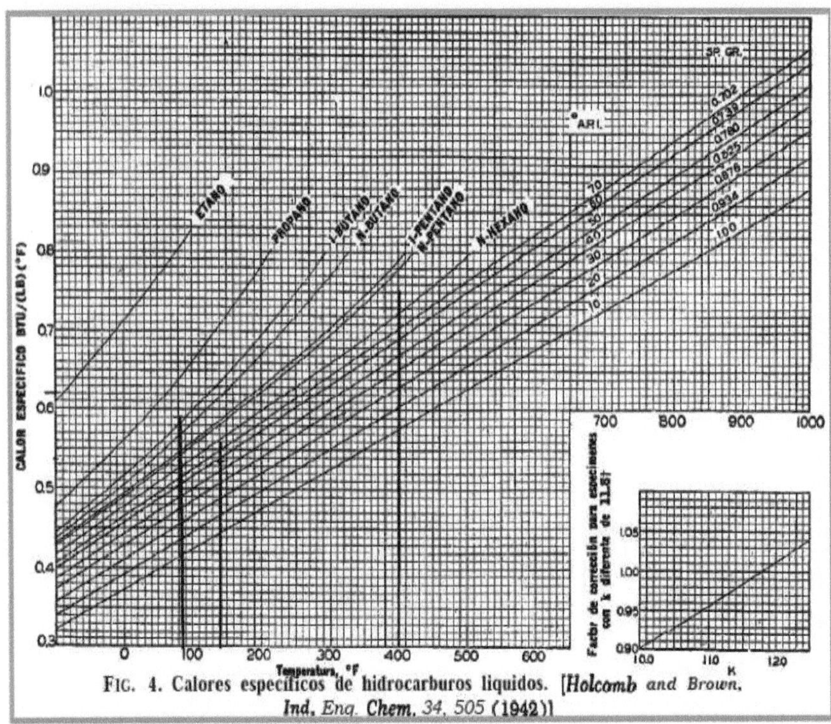

FIG. 4. Calores específicos de hidrocarburos líquidos. [Holcomb and Brown, Ind. Eng. Chem. 34, 505 (1942)]

Figure 2 **Plot of specific heats for liquid hydrocarbons as a function of temperature.** Source: Kern (1999)

Thermal conductivity "K" of pentane

Thermal conductivity is the measure of a material's ability to conduct heat, for pentane the thermal conductivity can be calculated by the following figure as a function of temperature.

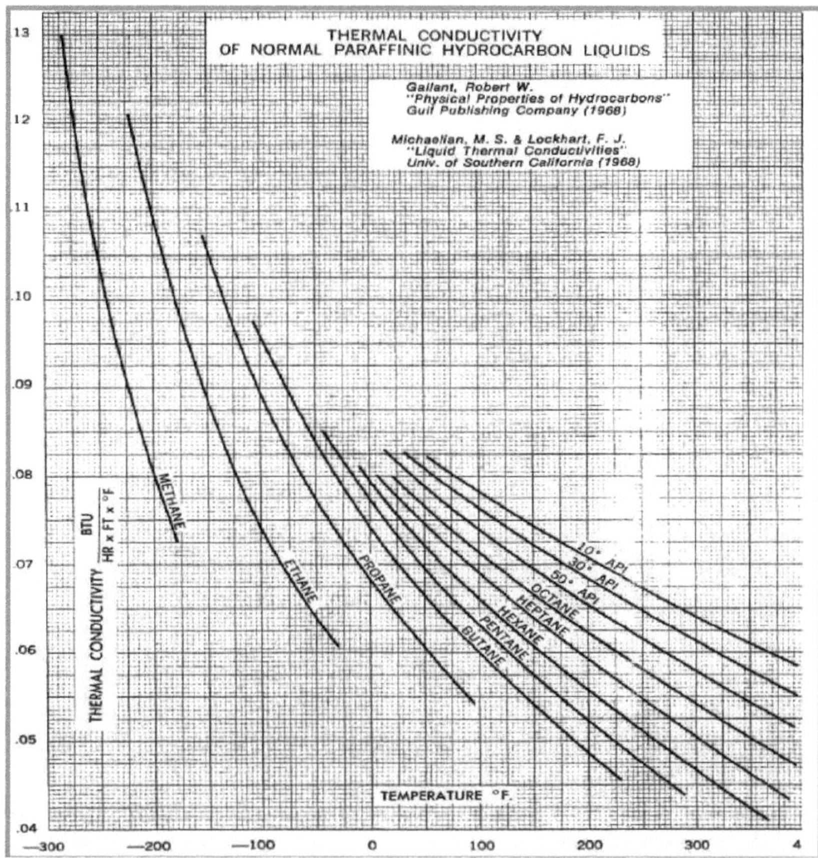

Figure 3. **Thermal conductivity K of some liquid hydrocarbons.** Source: TEMA (2004).

Absolute viscosity of pentane (p)

The viscosity of pentane can be determined by means of the following figure and is given in centipoise (cP) knowing the temperature, this is useful to calculate the Reynolds number and to know the flow regime, that is to say, if it is laminar, in transition or turbulent.

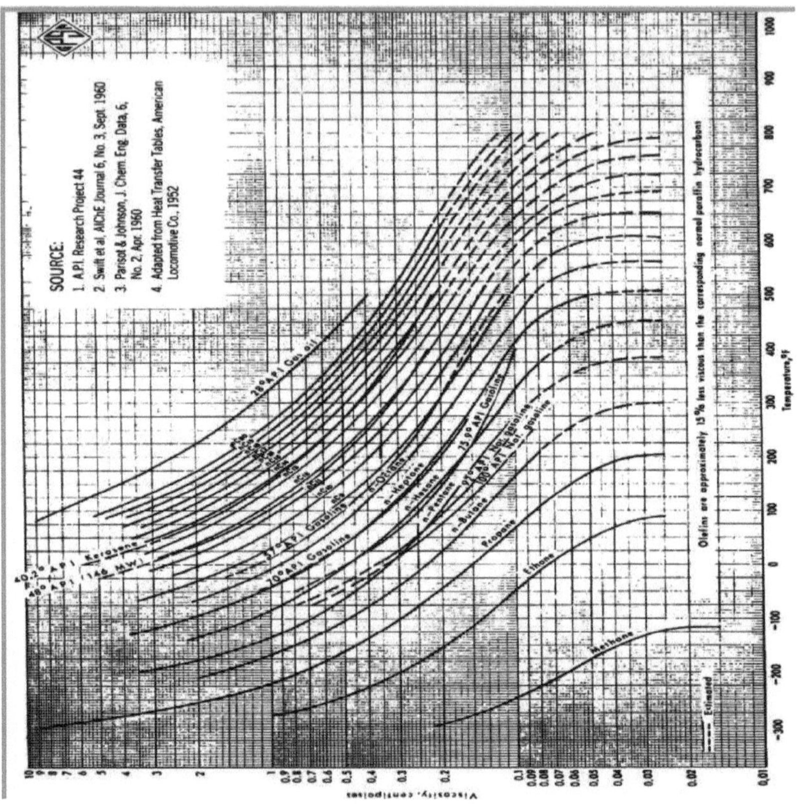

Figure 4. **Calculation diagram of dynamic viscosities (absolute) of some liquid hydrocarbons.**
Source: GPSA (2004): GPSA (2004).

Density of pentane

Density is defined as mass per unit volume. In general it depends on temperature and pressure, density for most gases is directly proportional to pressure and inversely proportional to temperature while for liquids it depends directly on pressure.

The density of pentane at different pressures and/or temperatures can be calculated by knowing the specific volume of the substance and the mass of the substance using the following formula:

$$\rho = m/V \; ; \; v = 1/v$$

Where **v** is the specific volume **m** is the mass **V** is the volume **p** is the density. Cengel (2009)

Another way to calculate the density is to locate the value of the pressure and temperature or two known values on the P-H diagram of the product and locate its specific volume and because the density is the inverse of the specific volume, then the unit (1) is divided by the value of the specific volume.

Propane

A by-product of natural gas processing and petroleum refining, its chemical symbol is **C3H8.** At ambient conditions propane is colorless and odorless. It becomes a liquid under moderate pressure and is stored and supplied in its liquid state. Propane is 270 times denser as a liquid than as a vapor and is an abundant energy resource. Some of the propane produced at the Bajo Grande fractionation plant is applied as a liquid refrigerant.

Physical and thermodynamic properties of propane

Many of the properties of propane, such as phase, density, specific volume, entropy, enthalpy as well as many other hydrocarbons can be seen on the pressure-enthalpy graph of propane. The value of a property can be known as a function of two known data. The left region of the graph is the saturated liquid region, the right region is the superheated vapor phase, while the area under the curve is the mixing region. This plot is useful because 49 the behavior of propane can be predicted under given conditions of pressure and temperatures or other known variable that when intercepted with another variable generates a value.

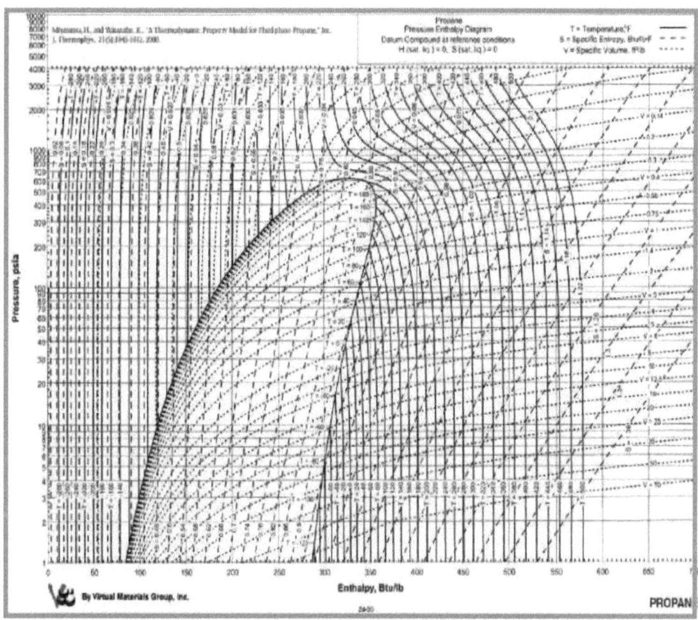

Figure 5. **Pressure-enthalpy diagram of propane**. Source GPSA 2004

Thermal conductivity "K" of propane

Thermal conductivity is the measure of the capacity of a material to conduct heat, for propane the thermal conductivity can be calculated by means of figure 3, locating the value of the temperature to which the fluid is subjected, following the line vertically upwards until intercepting it with the curve of the corresponding hydrocarbon, and then locating the value following the horizontal line to the left. This diagram is useful because at different temperatures the thermal conductivity of propane varies and it is necessary to know it.

Specific heat of propane (Cp)

Specific heat is defined as the energy required to raise the temperature of a unit mass of a substance by one degree. This value can be read as the specific heat of pentane in Figure 2.

Cooling system

Refrigeration systems are very common in industries involved in natural gas processing and processes related to petroleum refining, petrochemical and chemical industries.

Among these uses are the use of refrigeration systems for NGL and LPG recovery as well as for hydrocarbon dew point control, reflux condensation in fractionation towers and liquefied natural gas (LNG) plants.

The operation of a typical cooling system of a natural gas processing plant is generally determined by the use of heat exchangers, which are equipment composed of a casing or body and an internal arrangement of tubes or pipe segments to achieve thermal exchange between fluids.

The cooling fluid is passed through the side of the housing (1) (at low temperature) and the product to be cooled through the side of the tubes (A), the result is the exchange of temperature between fluids thus achieving the desired temperature. The selection of the cooling fluid depends on temperature requirements, availability, economy and previous experience.

Also, there is a container or vapor collection tank (economizer) (2) that is responsible for receiving the refrigerant, generally propane, which evaporates in the heat exchange and then these vapors are directed to a compressor (3) that puts the propane back to the operating pressure (4).

The propane is then passed through a condenser where it returns to its liquid phase (5), and is stored in a refrigerant storage drum, from where it is fed to the heat exchanger casings or chillers to start the process again. GPSA (2004)

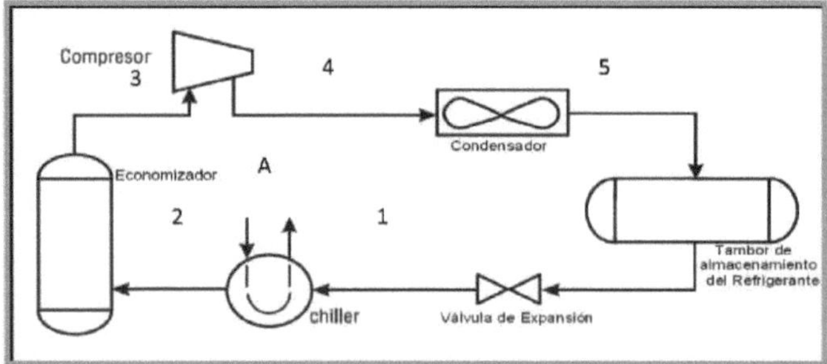

Figure 6. **Schematic diagram of a refrigeration system.** Source GPSA 2004

Refrigeration cycle.

The refrigeration effect can be achieved with the use of either of these two (2) cycles: Vapor expansion-compression and Absorption. The vapor compression refrigeration cycle can be sub-divided into four distinct steps using the pressure-enthalpy diagram, which are: Expansion, Evaporation, Compression and Condensation. The vapor compression refrigeration cycle is detailed below (Refer Figure 7). GPSA (2004)

- **Expansion Stage (first stage).**

The starting point for a refrigeration cycle is the availability of the liquid refrigerant. Point A, in Figure 7, represents the bubble point of the refrigerant liquid at its saturation pressure, **PA**, and enthalpy **HLA**. In the expansion stage the pressure and temperature of the liquid refrigerant are reduced, causing its passage through a control valve which allows reaching a lower pressure B, **PB**.

This lower pressure is determined based on the desired temperature of the refrigerant **TB**. It is worth mentioning that the enthalpy remains constant (insenthalpic process) since no work is produced, this effect is known as the "Joule-Thomsom effect".

At point B the enthalpy of the saturated liquid is **HLB** while the corresponding saturated vapor enthalpy is **HVB**. In this sense the total enthalpy of point A is equal to B. GPSA (2004)

- **Evaporation stage (second stage).**

The vapor formed in the expansion process (A-B), only corresponds to a fraction of the refrigerant, physically complete evaporation takes place inside an exchanger or a chiller as heat is absorbed in the exchange process producing evaporation. As shown in Figure 7 in the section (B-C) the temperature and pressure remain constant. While the enthalpy at point B is **HVB**. GPSA (2004)

- **Compression Stage (third stage).**

The gas phase refrigerant or refrigerant vapors, leave the chiller to move to the compression stage, in the same refrigerant is at a saturation pressure **PB** to which corresponds a temperature **TB** and an enthalpy **HVB**, entropy at this point is **Sc**. these vapors are compressed isotropically by the compressor raising its pressure to a pressure **PA** as can be seen in the section C-D' of Figure 7. GPSA (2004)

- **Condensation stage (fourth stage)**

The superheated vapor leaves the compressor at a pressure **PA**, and a temperature **TD**. In the condensation stage the superheated refrigerant is cooled to a constant pressure close to the dew point temperature **TA**.

Here the vapors begin to condense at a constant temperature. During the cooling and condensation process, all the heat and work absorbed by the refrigerant during the evaporation and compression process must be removed in order to complete the cycle and thus restart. GPSA (2004)

This cycle refers to the energy use since they are generally closed cycles that operate under the principles used in this process Expansion, Evaporation, Compression and Condensation, which generate high performance and efficiency thanks to the use of refrigerants that can be reused or regenerated in the process.

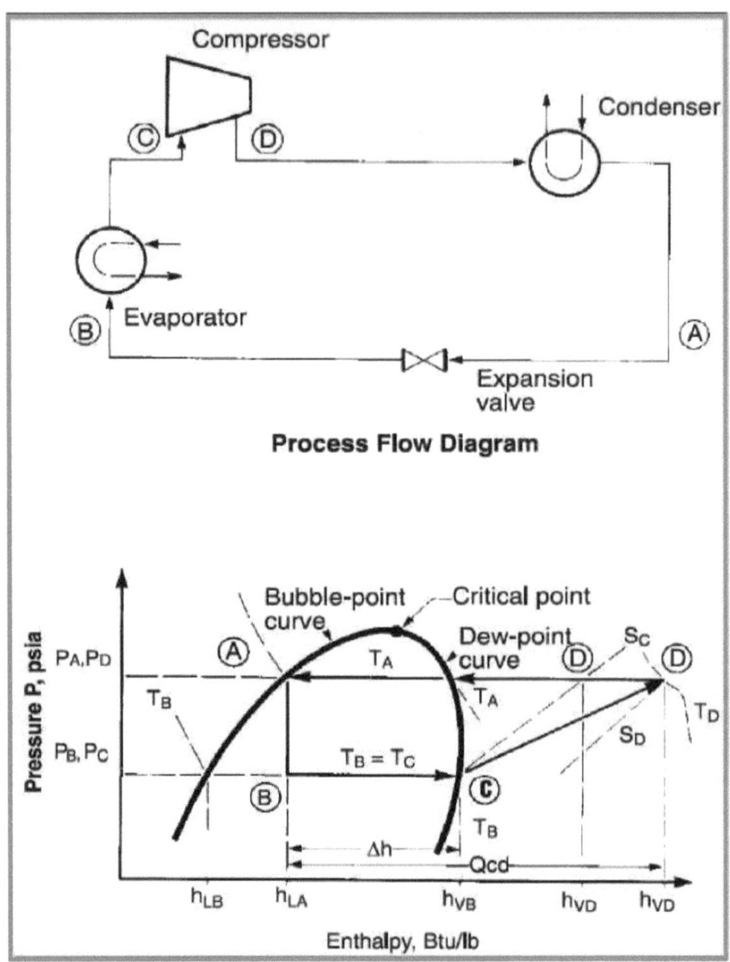

Figure 7. **Stages of the refrigeration cycle.** Source: GPSA 2004.

Most commonly used refrigerants.

The type of refrigerant to be used will generally depend on the desired temperature requirements, availability, economy and previous experience.

For example for a natural gas processing plant ethane and propane may be at hand, while in other plants not, in this sense they can be easily used due to their availability in plant. According to the GPSA the most common refrigerants are presented below with their physical properties:

Table 1

Most used refrigerants

ASHRAE Refrigerant Number	Chemical Name	Chemical Formula	Molecular Weight	Normal Boiling Point °F @ 14.696 psia	Critical Temperature °F	Critical Pressure psia	Freezing Point °F @ 14.696 psia	Liquid Viscosity Centipoise	Liquid Thermal Conductivity Btu (hr · sq ft · °F)/ft	Specific Heat Ratio $k = C_p/C_v$	Toxicity UL Group Classification
11	Trichloro-fluoromethane	OCl_3F	137.4	74.8	388.4	640.0	−168	0.421 @ NBT / 0.395 @ 86°F	0.0506 @ NBT / 0.0498 @ 86°F	1.13	5
114	Dichlorotetra-fluroethane	$CClF_2OClF_2$	170.0	38.4	294.3	474.0	−137	0.44 @ NBT / 0.32 @ 86°F	0.0405 @ NBT / 0.0366 @ 86°F	1.09	6
12	Dichlorodifluoro methane	OCl_2F_2	120.9	−21.6	233.6	597.0	−252	0.358 @ NBT / 0.206 @ 86°F	0.0518 @ NBT / 0.0392 @ 86°F	1.14	6
22	Chlorodifluoro methane	$CHClF_2$	86.5	−41.4	204.8	716.0	−256	0.33 @ NBT / 0.192 @ 86°F	0.0695 @ NBT / 0.0495 @ 86°F	1.18	5a
600	N-Butane	C_4H_{10}	58.1	31.1	305.6	550.7	−217	0.213 @ NBT / 0.159 @ 86°F	0.0663 @ NBT / 0.061 @ 86°F	1.09	5b
290	Propane	C_3H_8	44.1	−43.7	206.0	616.3	−305	0.21 @ NBT / 0.101 @ 86°F	0.076 @ NBT / 0.056 @ 86°F	1.14	5b
1270	Propylene	C_3H_6	42.1	−53.9	197.1	667.2	−301	0.15 @ NBT / 0.089 @ 86°F	0.082 @ NBT / 0.057 @ 86°F	1.15	5b
170	Ethane	C_2H_6	30.1	−127.4	9.01	707.8	−297	0.168 @ NBT / 0.039 @ 86°F	0.082 @ NBT / 0.048 @ 86°F	1.19	5b
1150	Ethylene	C_2H_4	28.1	−154.8	48.6	731.1	−272	0.17 @ NBT / 0.07 @ 86°F	0.111 @ NBT / 0.031 @ 86°F	1.24	5b
50	Methane	CH_4	16.0	−258.7	−116.7	667.8	−296	0.118 @ NBT	0.110 @ NBT	1.305	5b
717	Ammonia	NH_3	17.0	−28.0	270.4	1636.0	−108	0.25 @ 5°F / 0.207 @ 86°F	0.29 @ 32°F / 0.29 @ 86°F	1.29	2

Source: Data obtained from GPSA 2004. Mendoza and Barreto (2011)

Shell and tube type exchangers

The shell and tube exchanger consists of a parallel tube bundle enclosed in a cylindrical casing called a shell. This is the type of exchanger commonly used in refineries.

It can be designed for moderate to high pressures and to withstand external and internal stresses under normal operating conditions due to changes in temperature and pressure. Among the types of heat exchanger, the shell and tube type is generally used industrially, containing a large number of tubes (sometimes several hundred) packed in a parallel shell. Heat transfer takes place as one of the fluids moves inside the tubes, while the other moves outside the tubes through the shell.

Baffles are commonly placed to force the fluid to move in a transverse direction to the shell to improve heat transfer and also to maintain uniform tube spacing.

In this type of heat exchanger, the tubes open into certain large flow areas, called headers, which are located at both ends of the shell, and where the fluid on the tube side accumulates before entering and exiting the tubes.

There are three basic types of shell and tube heat exchangers, depending on the method used to keep the tubes inside the shell and the range of pressures from moderate to high:

- Fixed head exchanger.

The equipment has straight tubes, secured at both ends in tube plates welded to the casing. In this type of construction, it is sometimes necessary to incorporate an expansion joint or gasket joint in the casing due to differential expansion of the casing and tubes. This expansion is due to the operation of the equipment at different temperatures and the use of different materials in the construction.

The need for this gasket is determined by both the magnitude of the differential expansion and the expected operating cycle. When these seals or gaskets are not required, the equipment offers maximum protection against leakage of the fluid contained in the casing. They are used in services where the fluid in the casing is a clean fluid, such as water vapor, refrigerant, gases, certain types of cooling water, etc.

- Heat exchanger with U-tubes.

It uses U-shaped tubes, with both ends of the tubes clamped to a single tube plate, thus eliminating differential expansion problems because the tubes can expand and contract freely, the U-shape absorbing these changes.

- Fixed head exchanger with expansion joint or gasketed joint.

It features two tube plates, but with only one of them welded to the shell and the other moving freely, thus avoiding differential expansion problems. Shell-tube heat exchangers are further classified according to the number of passes through the shell and through the tubes. For example, in heat exchangers where the tubes form a U-shape in the shell, they are said to have one pass through the shell and two passes through the tubes (a). Similarly, an exchanger comprising two passes in the shell and four passes in the tubes is called two passes through the shell and four passes through the tube (Cengel 2nd Edition).

General design considerations for an Exchanger

The rate of heat transfer from one fluid to another through a metal wall is proportional to the overall heat transfer coefficient, the wall area and the temperature difference between the hot and cold fluid:

$$Q = U_o \times A \times DTMe$$

Where:

Q= Heat transfer rate.

Uo= Overall heat transfer coefficient based on the external surface area of the metal.

A= External area of the metal surface through which heat transfer occurs.

DTMe=Difference of logarithmic mean temperatures between hot and cold fluids.

When specifying a heat exchanger, the designer almost always knows, or can calculate without much difficulty, the Q and DTMe terms for the given process conditions. To obtain the appropriate value of the required heat transfer area, only the Uo coefficient needs to be evaluated.

Total heat transfer coefficient (U)

Commonly a heat exchanger is related to two flowing fluids separated by a solid wall as explained above. First in the exchange process, heat is transferred from the fluid to the tube wall by convection, then through the wall by conduction and finally from the wall to the cold fluid by convection. The calculation of the total heat transfer coefficient "U" is calculated by the following formula:

$$U = \cfrac{1}{\left(\dfrac{1}{hi} \times \dfrac{Ao}{Ai}\right) + \left(Rfi \times \dfrac{Ao}{Ai}\right) + Rw + \dfrac{1}{ho} + Rfo}$$

Where **Rw** is the pipe wall resistance, Rfi and Rfo are the internal and external fouling factor of the pipes.

Fouling Factor (Fouling Factor):

The performance of heat exchangers often deteriorates over time as a result of the accumulation of deposits on the heat transfer surface.

The layer of deposits represents additional resistance and is caused to slow down. The net effect of these accumulations on heat transfer is represented by a scale factor **Rf**, which is a measure of the thermal resistance introduced by the scale. In applications where it is likely to occur it should be considered in the design and selection of heat exchangers.

The total heat transfer coefficient ratio given above (4) is valid for clean surfaces and needs to be modified to take into account the effects of fouling on the inner and outer tube surfaces for a shell and tube heat exchanger can be expressed (Cengel "Heat Transfer 2nd Edition").

Operating Temperature

It is the process fluid temperature expected for normal operation. The operating temperatures of an exchanger are set by the process conditions.

However, in certain cases, the exchanger designer can establish the operating conditions, but in no case can these be less than the minimum required by the process.

$$R = R_i + (R_{f,\,i} / A_i) + R_{pared} + (R_{f,\,o} / A_o) + R_o$$

Design Temperature

It is the temperature of the metal that represents the most severe coincident temperature conditions. This temperature is used for mechanical design of equipment and piping, including material selection.

This design temperature should be at least 10°C (18°F) higher than the maximum operating temperature, but in no case lower than the maximum temperature in cases of emergency, such as utility failure, operation lockout, instrument failure, etc. The design temperature of equipment and systems protected by relief valves should be at least the maximum temperature coincident with the set pressure of the respective relief valve. Process Design Manual PDVSA (1995)

The design temperatures of the hot and cold sides of an exchanger are determined independently based on process considerations and typically using the following criteria:

- For heat exchangers operating at temperatures between 0°C (32°F) and 399°C (750°F), the design temperature should be defined as the maximum expected operating temperature plus 14°C (25°F).

- The minimum design temperature should be 66°C (150°F) for exchangers operating above 0°C (32°F).

- For exchangers operating at 0°C (32°F) and below, the design temperature should be defined as the minimum expected operating temperature.

Where there is a possibility of exposing the tubes, tube sheet and floating head to the inlet temperature of the hot fluid as a result of loss of the cooling medium, these components should be designed for the maximum expected operating temperature of the hot fluid.

Design pressure

It is the maximum pressure, internal or external, to be used to determine the minimum thickness of pipes, vessels or other equipment. For partial or total vacuum conditions, the external pressure is the maximum pressure difference between atmospheric and internal pressure existing in the equipment. Unless otherwise specified, the design pressure is the pressure specified at the top of the vessel. PDVSA Process Design Manual (1995). The design pressures of the hot and cold sides of an exchanger are determined independently based on the operating conditions, according to the following criteria:

- The design pressure should be equal to the maximum expected operating pressure plus the greater of 10% of that pressure and 172 kPa man.

- The minimum design pressure should be 30 psig (207 kPa man.).

Pressure drop

The pressure drop in an exchanger is a product of three types of losses: friction losses due to flow, losses due to changes in flow direction and losses caused by expansion and contraction at the inlets and outlets of nozzles and tubes. The method for calculating the pressure drop is different for each type of heat exchanger.

Intercambiadores de Carcaza y Tubos, Doble Tubo y Enfriadores de Aire		
Gases y Vapores (Alta Presión)	35–70 kPa	5–10 psi
Gases y Vapores (Baja Presión)	15–35 kPa	2–5 psi
Gases y Vapores (Presión Atmosférica)	3.5–14 kPa	0.5–2 psi
Vapores (Vacío)	< 3.5 kPa	< 0.5 psi
Vapores (Condensadores de Torre de Vacío)	0.4–1.6 kPa	3–12 mm HG
Líquidos	70–170 kPa	10–25 psi

Figure 8. **Typical pressure drops of shell and tube heat exchangers.** Source PDVSA (1995)

Arrangement of currents

Multiple shell and tube heat exchangers are usually connected in series because of the advantage gained in the effective temperature difference, DTMe.

Number of housings

The total number of shells required for an exchanger is generally set by the magnitude of the difference that exists between the outlet temperature of the hot fluid and the outlet temperature of the other fluid. This difference is known as the "extent of the temperature crossover". The "crossover" plus other variables determine the value of Fn, called the temperature correction factor; this factor must always be equal to or greater than 0.80. (The value of Fn decreases slowly between 1.00 and 0.80, but then decreases rapidly until it reaches a value close to zero.

The maximum area per shell is usually limited to a unit with a shell internal diameter of 1219 mm (48 in.) or a tube bundle of 13.6 t (15 short tons). (These limitations do not necessarily apply to fixed tube sheet heat exchangers).

In special cases, such as reboilers and fixed tube plate exchangers, large areas per shell are occasionally used (areas up to 2300 m2 (25000 ft2) have been used in fixed tube plate exchangers).

Tube Selection

The nominal tube size of a heat exchanger is the outside diameter in inches, typical values are

5/8, % and 1 in. With lengths of 8, 10, 12, 16 and 20 ft. Typical lengths are 16 ft.

Tube thicknesses are given according to BWG (Birmingham Wire Gauge) and are determined by working pressure and fouling factor.

Typical values are 16 or 18 for Admiralty brass and 12, 13 or 14 for carbon steel. Tube configuration can be square, square turned 90°, or triangular.

The square is used for ease of mechanical cleaning (TEMA standard). According to the TEMA standard a first approximation of tubes to be used is:

Diameter % in, triangular arrangement and spacing between tubes 1 in with 16 ft long and thickness 14 BWG.

Material	Composición	Conductividad Térmica, K	
		W/m°C	BTU/hpie²°F/pie
Admiralty	(71 Cu – 28 Zn – 1 Sn)	111	64
Acero inoxidable tipo 31	(17 Cr – 12 Ni – 2 Mo)	16	9
Acero inoxidable tipo 304	(18 Cr – 8 Ni)	16	9
Aluminio		202	117
Acero al Carbono		45	26

Figure 9. **Thermal conductivities of tubes, for heat exchangers.** Source PDVSA (1995)

de = DE del tubo		dw = Espesor de pared		di = DI del tubo		área Interna		Superficie Externa por metro de longitud	por pie longitud
mm	pulg	mm	pulg	mm	pulg	mm²	pulg²	m cuadrado	pie cuadrado
19.05	3/4	2.77	0.109	13.51	0.532	143.8	0.223	0.0598	0.1963
19.05	3/4	2.11(1)	0.083(1)	14.83	0.584	172.9	0.268	0.0598	0.1963
19.05	3/4	1.65(2)	0.065(2)	15.75	0.620	194.8	0.302	0.0598	0.1963
19.05	3/4	1.24	0.049	16.56	0.652	215.5	0.334	0.0598	0.1963
25.40	1	3.40	0.134	18.59	0.732	271.6	0.421	0.0798	0.2618
25.40	1	2.77(1)	0.109(1)	19.86	0.782	309.0	0.479	0.0798	0.2618
25.40	1	2.11(2)	0.083(2)	21.18	0.834	352.3	0.546	0.0798	0.2618
25.40	1	1.65	0.065	22.10	0.870	383.2	0.594	0.0798	0.2618
38.10	1 1/2	3.40	0.134	31.29	1.232	769.0	1.192	0.1197	0.3927
38.10	1 1/2	2.77	0.109	32.56	1.282	832.9	1.291	0.1197	0.3927
38.10	1 1/2	2.11	0.083	33.88	1.334	901.3	1.397	0.1197	0.3927

NOTAS:

(1) Espesor de pared preferido para tubos de acero al carbono.
(2) Espesor de pared preferido para tubos de aleaciones de cobre.

ESCALA EQUIVALENTE

mm	pulg	BWG
4.19	0.0165	8
3.76	0.148	9
3.40	0.134	10
3.05	0.120	11
2.77	0.109	12
2.41	0.095	13
2.11	0.083	14
1.83	0.072	15
1.65	0.065	16
1.47	0.058	17
1.24	0.049	18

Figure 10. **Tube data for heat exchangers.** Source: PDVSA (1995): PDVSA (1995)

Pipe fittings

There are four types of tube arrangements with respect to the transverse direction between the edges of the baffles on the casing side: square (90°), rotated square (45°), triangular (30°) and rotated triangular (60°) (Tube Layout).

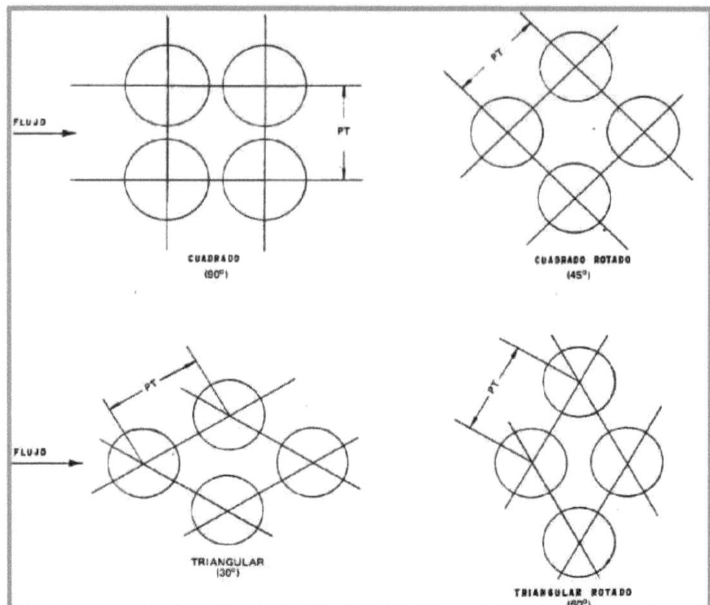

Figure 11. **Tube arrangements for heat exchangers.** Source PDVSA (1995)

Deflectors

The path of the fluid contained in the shell depends on the type and arrangement of the baffles. In some cases the flow pattern affects the heat transfer considerably, while in others it is unimportant, for example in heat exchangers with condensation or when the value of the heat transfer coefficient on the shell side considerably exceeds the corresponding one on the tube side. Most baffles serve two functions: (1) to direct the flow along the desired path and (2) to support the tubes and prevent vibration. The most common types of baffles are: transverse, longitudinal and window baffles, which are described below (PDVSA Standard MDP-05-E-01).

Crossflow Baffles (Crossflow Baffles)

Transverse baffles support the tubes, restrict the vibration of the tubes by collision with the fluid and direct the flow, on the casing side, transversely to the tube bundle (perpendicular to the tube axis); providing a high degree of turbulence and a high heat transfer coefficient, but increasing the pressure drop. Within this type of baffles, the most known and efficient is the segment baffle (PDVSA Standard MDP-05-E-01).

Cutting of the deflector

It is the portion of the baffle "cut" to allow flow through the baffle. The size of this cutout affects the heat transfer coefficient and pressure drop. For segment baffles, this portion is expressed as a percentage and is given as the ratio of the height of the cutout to the casing diameter. Usually, segment baffle cutouts are approximately 25% of their area, although the maximum practical cutout for tube supports is approximately 48%.

Work performed (single-stage) by Heat Transfer Research Inst. (HTRI) on segment baffles indicates that the optimum shear range is 15 to 30%, with 25%. as the optimum. A higher cutoff would result in poor flow distribution with dead spaces behind the baffle and decreased pressure drop and transfer coefficient. A lower cutoff would result in a high fluid velocity in the cutoff with increased pressure drop, and there would also be dead spaces and eddies behind the baffle. (PDVSA Standard MDP-05-E-01).

Deflector spacing

It is the longitudinal spacing between baffles to the nearest 1/4" (6 mm). The smaller the baffle spacing, the higher the transfer coefficient and the higher the pressure drop; therefore, this distance should be defined in such a way as to allow a high velocity and a high transfer coefficient within the allowable pressure drop limits, i.e., for optimum heat exchanger design. The maximum baffle spacing should not exceed the shell diameter and should be adequate to provide support for the tubes and prevent possible vibration of the tubes.

This dimension is defined in the TEMA as "maximum unsoported span"; the recommended values being a function of the pipe size and, for flows without phase change, the casing diameter.

If there is no phase change in the fluid on the casing side, the baffle spacing should not exceed the inside diameter of the casing; otherwise the fluid would have to flow parallel to the tubes, instead of perpendicular, thus producing a much lower heat transfer coefficient. When condensation or vaporization is present, the maximum baffle spacing is only a function of the tube diameter (PDVSA Standard MDP-05-E-01).

Minimum baffle spacing

Required to maintain good flow distribution, it is 20% of the inside diameter of the casing but not less than 50 mm (2 in.). Too small a baffle spacing forces the fluid in the casing to deflect, thus producing a decrease in the heat transfer coefficient. The orientation of the baffle cutouts depends on the type of fluid, pipe arrangement and service (PDVSA Standard MDP-05-E-01).

Longitudinal baffles

These types of baffles are used to divide the casing into two or more sections, creating multi-pass casings. They should be used welded to the casing and tube sheet to prevent fluid from leaking through the collapse between the baffle and the casing, which would decrease the efficiency of the heat transfer operation. Due to the difficulties encountered in performing a good welding process, when multipass casings are required, it is more economical to use separate casings for each pass; unless the casing diameter is large enough to allow easy welding of the baffle to the casing. (PDVSA Standard MDP- 05-E-01).

Window deflectors

When low shell pressure drop is required in an exchanger, for example in gas handling, the use of cross-flow baffles is impractical. In this case parallel flow baffles, known as window baffles, should be used. The main function of window baffles is to support the tubes while allowing the fluid to flow parallel to the tubes. Within this type of baffles the best known and most efficient is the segment baffle (PDVSA Standard MDP-05-E-01).

Analysis of heat exchangers.

The two methods for the analysis of heat exchangers are: the "Logarithmic Mean Temperature

Difference (LMTD)" which is applied when the inlet and outlet temperatures of the fluid and the mass flow rate are known. The other method is the "NTU-Effectiveness Method" which is used to predict the hot and cold fluid temperatures in a specific heat exchanger.

For either of the two methods used, the following considerations must be taken into account:

- Heat exchangers tend to operate for long periods of time without changes in their operating conditions and can therefore be considered steady flow devices. This gives way to the next consideration.

The mass flow rate (mass flow) of each fluid remains constant and the properties of the fluids, such as temperature and velocity, at any given inlet or outlet remain the same.

- Also, the fluid streams experience little or no change in their velocities and elevations and, as a consequence, changes in kinetic and potential energy are negligible.
- In general, the specific heat of a fluid changes with temperature; but over a specific range of temperatures, it can be considered constant at some average value, with little loss of accuracy.
- Axial heat conduction along the tube is usually negligible and can be considered negligible.
- Finally, it is assumed that the outer surface of the heat exchanger is perfectly insulated, so that there is no heat loss to the surrounding medium, and any heat transfer occurs between the fluids.
- Heat is transferred from the hot fluid to the cold fluid according to the second law of thermodynamics.

With these assumptions, the first law of thermodynamics requires that the rate of heat transfer (amount of heat transferred per unit time) from the hot fluid is equal to the heat transfer to the cold fluid, i.e., the rate of heat transfer to the cold fluid is equal to the rate of heat transfer from the hot fluid to the cold fluid, i.e., the rate of heat transfer to the cold fluid is equal to the rate of heat transfer to the cold fluid:

$$\dot{Q} = \dot{m}_c \cdot C_{pc} \left(T_{c,\,Ent} - T_{c,\,Sal} \right)$$

Y

$$\dot{Q} = \dot{m}_F \cdot C_{pF} \left(T_{F,\,sal} - T_{F,\,Ent} \right)$$

Where the subscripts C and F refer to the hot and cold fluids respectively, Q is the amount of heat given up or absorbed per unit time, "m" mass flow per unit time, Cp specific heat, *Tent* inlet temperatures and *Tsal* outlet temperatures. In the analysis of heat exchangers, it is often convenient to combine the product of the mass flow and the specific heat of a fluid into a single quantity and this is called the Heat Capacity Ratio, and is defined for the hot and cold fluid streams as:

This represents the rate of heat transfer required to change the temperature of that stream by 1 °C as it flows through the heat exchanger. And it can be noted that in a heat exchanger a fluid with a large heat capacity ratio will experience a small change in temperature and vice versa. Therefore doubling the mass flow of a fluid while leaving everything else unchanged will halve the temperature change of that fluid.

$$\dot{Q} = C_c \, (T_{c,\,Ent} - T_{c,\,Sal}) \quad ; \quad \dot{Q} = C_F \, (T_{F,\,sal} - T_{F,\,Ent})$$

The rate of heat transfer in a heat exchanger can be expressed analogously to Newton's law of cooling as:

$$C_c = \dot{m}_c \cdot C_{pc} \quad y \quad C_F = \dot{m}_F \cdot C_{pF}$$

$$\dot{Q} = U \, A_s \, \Delta T_m$$

Where "As" is the heat transfer area (Cengel "Heat Transfer 2nd Edition").

Logarithmic Mean Temperature Difference (LMTD) method.

For this method, a parallel flow exchanger, a counter flow exchanger and a cross flow exchanger are considered. The first consideration is made for the parallel flow exchanger.

Parallel flow exchanger:

In the parallel flow heat exchanger the temperature difference between the hot and cold fluids is large at the inlet of the heat exchanger as shown in Figure 14, but decreases exponentially towards the outlet, in this case the temperature of the cold fluid decreases and the hot fluid increases along the heat exchanger, but the temperature of the cold fluid can never exceed that of the hot fluid no matter how long the heat exchanger is.

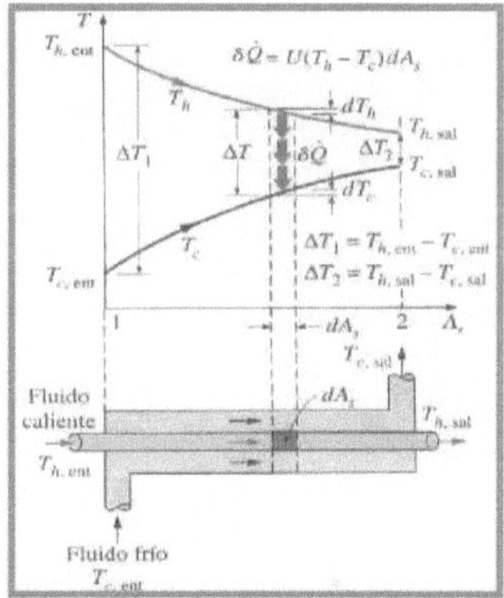

Figure 12. **Heat transfer in a parallel flow heat exchanger.** Source Cengel (2009)

Counterflow heat exchanger

Considerations for counterflow heat exchangers are as follows:

$$\Delta T_{ml} = F \; \Delta T_{ml,\,CF}$$

- In counter flow heat exchangers the fluids enter at opposite ends as shown in Figure 15, unlike the parallel flow heat exchanger, in counter flow the outlet temperature of the cold fluid may exceed the outlet temperature of the hot fluid.

- In the limiting case, the cold fluid will heat up to the inlet temperature of the hot fluid, however the outlet temperature of the cold fluid can never be higher than the inlet temperature of the hot fluid as this would be a violation of the second law of thermodynamics.

- For specific inlet and outlet temperatures, the logarithmic mean temperature difference is always larger than that of a parallel flow heat exchanger, and thus a smaller surface area (and therefore a smaller heat exchanger) is needed to achieve a specific heat transfer rate. It is therefore common practice in heat exchangers to use counter flow arrangements.

$$P = \frac{t_2 - t_1}{T_1 - t_1} \qquad\qquad R = \frac{T_1 - T_2}{t_2 - t_1}$$

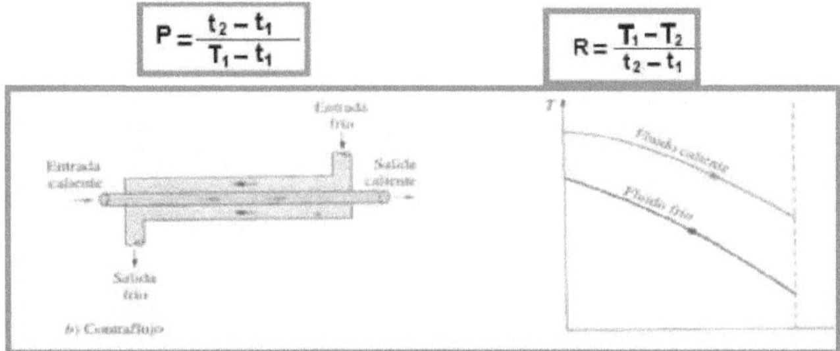

Figure 13. **Heat transfer in a counter flow heat exchanger.** Source: Cengel (2009)

Cross-flow heat exchangers

The relationship for the logarithmic mean temperature difference is only developed for counterflow and crossflow heat exchangers. Similar relationships are also developed for cross-flow heat exchangers but the resulting expressions are too complicated due to the complex flow conditions. In these cases it is convenient to relate the equivalent temperature difference to the ratio of the logarithmic mean difference for the counterflow case as:

Where F is the **correction factor** which depends on the geometrical configuration of the exchanger, and the inlet and outlet temperatures of the hot and cold fluid streams. $\acute{A}T_{ml}$, **CF** is the mean logarithmic temperature difference, for the case of the counterflow heat exchanger, with the same inlet and outlet temperatures and is determined based on equation (12). The following tables give the values of F for common configurations of cross-flow shell and tube heat exchangers, as a function of the ratios P and R, between two temperatures, defined as:

Where P and R are the ratios between two temperatures (i.e. it relates the temperatures of the casing and the tubes).

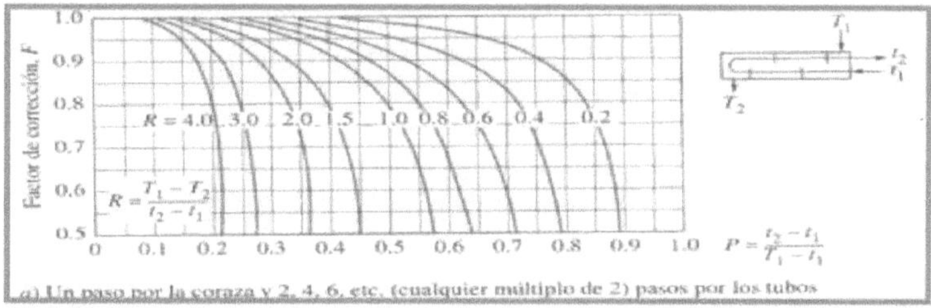

Figure 14. **Correction factor for heat exchangers**. Data obtained in Cengel Heat Transfer. Source: Mendoza and Barreto

Logarithmic Mean Temperature Difference.

As mentioned above the temperature difference between the hot and cold fluids varies along the heat exchanger and it is convenient to have an average temperature difference ÁTm, for the two types of heat exchangers mentioned above. This relationship can be applied to both since the exact value is obtained.

$$\Delta T_{ml} = \frac{\Delta T_1 - \Delta T_2}{Ln(\Delta T_1 / \Delta T_2)}$$

This is the logarithmic average temperature difference and is the appropriate form of the average temperature difference to use in heat exchanger analysis. In this case **ATI** and **AT2** represent the temperature difference between the two fluids at both ends (inlet and outlet) of the heat exchanger. It should be noted that it makes no difference which end is assigned as the inlet or outlet.

The logarithmic mean temperature difference is obtained by following the actual profile of fluid temperatures along the heat exchanger and is an accurate representation of the average temperature difference between the hot and cold fluids, and reflects without margin of error the exponential decay of the local temperature difference.

Effectiveness-NTU Method

After discussing the logarithmic mean temperature difference method, and concluding that it is the most appropriate method for sizing and selecting a heat exchanger when fluid inlet and outlet temperatures and fluid mass flow rates are known or can be determined. A second class of problem encountered in heat exchanger analysis is the determination of hot and cold fluid heat transfer rates and fluid outlet temperatures for prescribed values of mass flow and inlet temperatures when specifying the type and size of heat exchangers.

In this case the surface area "*A*" for heat transfer is known but the outlet temperatures are ignored. The task in this type of problem is to determine the performance with respect to heat transfer of a specific heat exchanger, or to determine whether an exchanger that has in physical can do the job. The **Nomenclature** method **for the identification of heat exchangers according to TEMA standard**

A TEMA-compliant shell and tube heat exchanger is identified by three letters, the first letter indicating the type of the stationary head. Types **A** (Channel and removable cover) and **B** (Shell) are the most common.

The second letter indicates the type of hull. The most common is **E** (one-pitch hull), the two-pitch

F is more complicated to maintain. Types **G, H** and **J** are used to reduce pressure losses in the hull. Type K is the type of boiler reboiler used in fractionation tower.

The third letter indicates the type of head at the rear end, type **S, T** and **U** are the most commonly used. The S type (floating head with support device) the diameter of the head is larger than the hull diameter and needs to be disassembled to remove it. The **T type** (floating head without counterflange) can be removed without disassembly, but requires a larger hull diameter for the same exchange surface. The **U-type** (U-tube bundle) is the most economical.

Types of Shell & Tube Exchangers and their parts:

Internal Floating Head Exchanger (AES type)

It is the most common model, it has a single pitch hull, double pitch tubes with removable channel and cover, floating head with support device. It has transverse diverters and support plates. Its characteristics are:

- It allows the thermal expansion of the tubes with respect to the hull.
- Allows disassembly

- Instead of two steps it can have 4, 6 or 8 steps.
- The transverse diverters, with the percentage of pitch and their spacing, modify the speed in the hull and its load loss.
- The flow is countercurrent and downstream in the middle of the tubes.

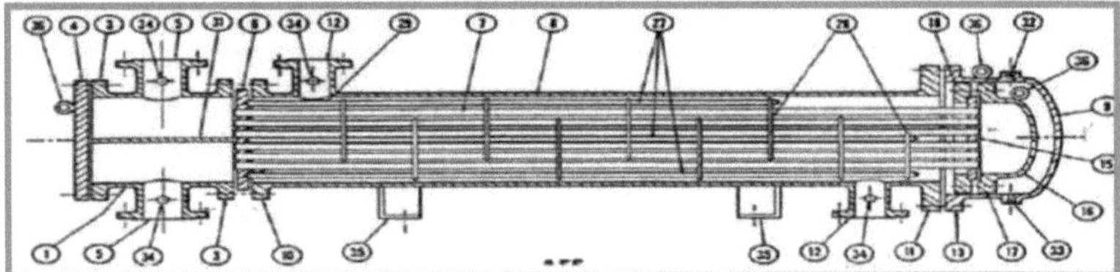

Figure 15. **AES type heat exchanger.** Source: PDVSA GAS: PDVSA GAS.

Exchanger type AEP

This model allows some movement of the floating head and can be disassembled for cleaning. It has the disadvantage of requiring more maintenance to maintain the packing and prevent leakage.

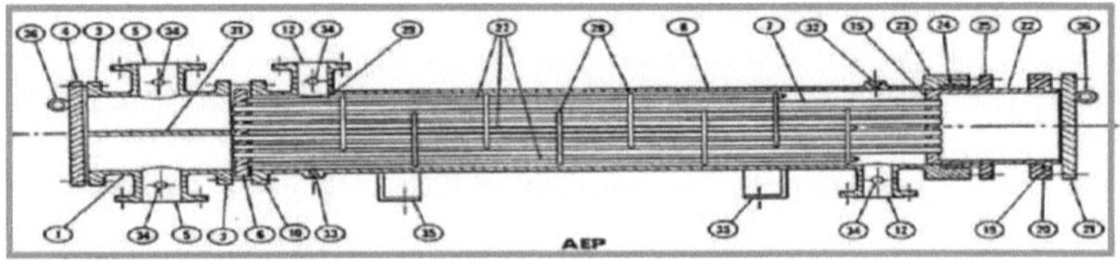

Figure 16. **AEP type exchanger.** Source: PDVSA GAS: PDVSA GAS.

Integrated Head and Tube Exchanger (CFU type)

This model has a U-tube assembly which allows for easy disassembly of the tube assembly. It has the disadvantage of replacing a damaged tube. It has the center baffle attached to the tube plate.

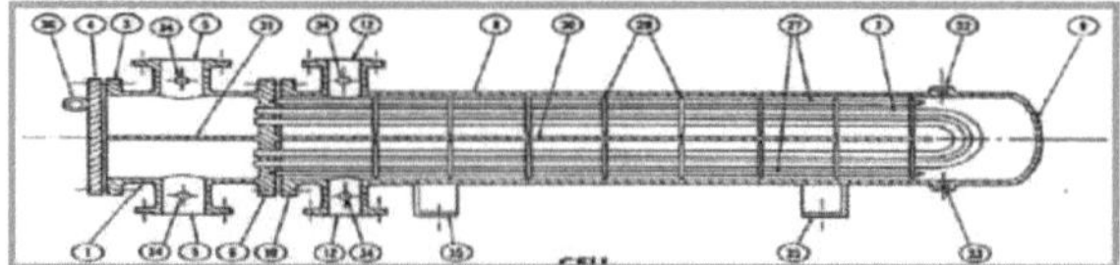

Figure 17. **CFU type exchanger**. Source: PDVSA GAS: PDVSA GAS.

Split Flux Capacitor (type AJW)

It is mainly used to condense vapors, as it reduces the pressure drop. Part of the exchanger is used as a condenser and part can be used as a cooler. The central diverter splits the flow in two and the rest of the diverters carry it through the tubes for cooling.

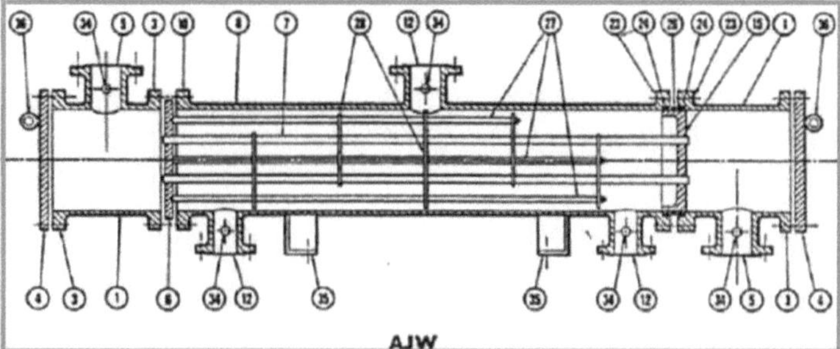

Figure 18. **AJW type heat exchanger**. Source: PDVSA GAS.

Exchanger type AKT

This exchanger is characterized by the shell configuration. The tube bundle can also be A-U, giving rise to the AKU.

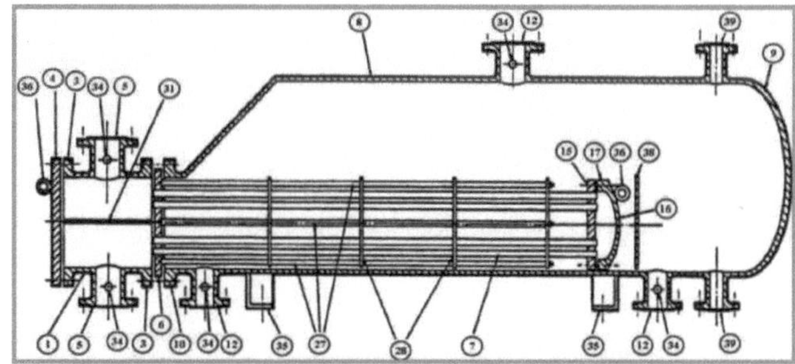

Figure 18 **AKT type heat exchanger**. Source: PDVSA GAS.

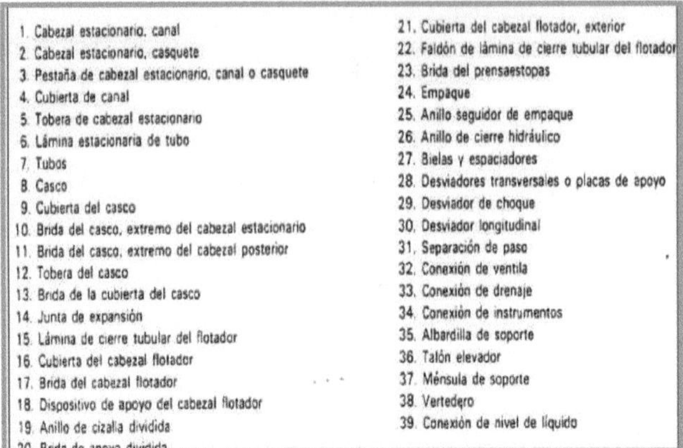

Figure 19 **Parts of a Heat Exchanger.** Source: PDVSA GAS de Occidente: Data obtained by reviewing PDVSA GAS de Occidente Management files. Mendoza and Barreto (2011)

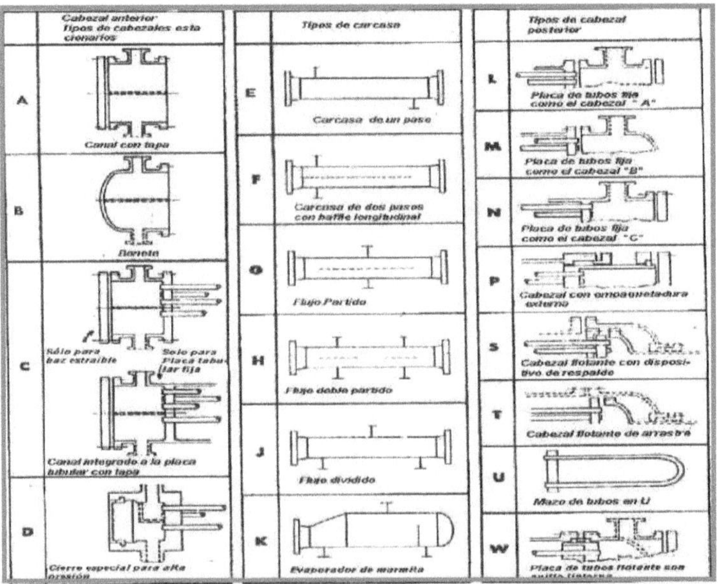

Figure 21. **Nomenclature TOPIC "parts of a shell and tube heat exchanger".** Source: Cao (2004)

General guide for heat exchanger design

The PDVSA MDP-05-E-01 Rev.0 standard suggests the methodology to be used as a general guide for the design of any type of heat exchanger.

The methodology to be used according to the PDVSA standard is mentioned below:

Step 1

Obtaining process (e.g., properties of the currents) and mechanical (e.g., tube arrangement) information and the function of the unit or service within the process (e.g., condenser). According to the procedures presented previously, it is required to obtain the following minimum process information:

1. Phase and nature of flow: liquid, vapor, gas, two-phase, water vapor, water, hydrocarbons, etc.
2. Total flows (entering and leaving the unit, minimum on one side): mass or volumetric, specified by phase in the case of two-phase flow.
3. Fluid properties: specific heat, latent heat, thermal conductivity, viscosity, molecular weight or specific gravity or density.
4. Operating conditions: temperature and pressure at the inlet and/or outlet of the unit
5. Permissible pressure drops: maximum allowable on each side
6. Fouling factors: preferably based on operational experience.
7. Design conditions: temperature and pressure on each side of the unit. Certain details, such as pipe length, material type specifications, limitations on casing dimensions, etc., are usually specified by the customer.

Step 2.

Definition of the type of exchanger according to the selection criteria presented.

Step 3

Location of typical design criteria for the service in question, additional criteria and considerations, and the configuration of the exchanger. **Step 4**

Thermal and hydraulic design of the heat exchanger using the following calculation procedure:

1. Calculate Q from process considerations.
2. Calculate DTMe from process considerations, exchanger type and tube arrangement.
3. Assume the value of the overall heat transfer coefficient Uo.
4. Calculate the area A based on the assumed Uo.
5. From the calculated A, determine the physical dimensions of the heat exchanger.
6. Calculate the pressure drop across the exchanger and modify the internal design, if required, to obtain a reasonable balance between pressure drop and exchanger size.
7. Calculate Uo from the physical properties of the fluids, fouling factors and the exchanger arrangement.
8. Recalculate A based on Q and the calculated values of Uo and DTMe.
9. Compare the calculated A with the assumed A and repeat the calculations until these areas are equal.

Step 5

Definition and sizing of inlet and outlet nozzles

Primary Elements of Level Control

In modern times, particularly since the industrial revolution, the need for measurement and control has kept pace with the increasing number of variables, conditions and factors that are involved in industrialization. However, the number of variables that must be measured and controlled is always uncountable. Throughout the oil and gas industry, there has been a need to automate a large number of processes, giving way to instrumentation, to control all the processes that are of vital importance. There are four basic variables of interest: temperature, pressure, level and flow. PDVSA Instrumentation Technical Report (2010).

Temperature.

It is the amount of heat expressed in degrees contained in a body. In other words, temperature measures or gives us an idea of the degree of heat or cold contained in a body, heat being one of the ways of presenting energy.

Temperature measurement.

Temperature measurement is one of the most common and important measurements in industrial processes. The limitations of the measurement system are defined in each type of application, by the accuracy, by the speed of temperature acquisition, by the distance between the measuring element and the receiving device and by the type of indicating instrument, recorder or controller required. An important characteristic of temperature in any industrial process is that it can be used to control many process parameters. PDVSA Technical Report (2010).

Temperature Measuring Instruments.

Searching for new measurement principles and improving existing methods to meet the ever-increasing demands of the industry. To cope with the demands of this extensive spectrum for measuring temperature, several physical principles have been considered as the basis for temperature measuring instruments. The range of an instrument to be considered when looking for a solution to a problem. Sensitivity, accuracy, speed of response, service life, cost, availability of various automatic control models and resistance to corrosion, vibration and other conditions for industrial use, are just some of the factors that require attention when selecting equipment to measure temperature.

There are several temperature sensors used in the industry, among them are: glass thermometers, bimetal thermometers, resistance thermometers for very cold temperatures, thermocouples, thermistors, radiation pyrometers, among others. PDVSA Instrumentation Technical Report (2010).

Pressure.

It is a physical quantity that measures the force per unit area, and serves to characterize how a given resultant force is applied on a surface.

Pressure is one of the most important industrial variables since chemical reactions within a reactor occur at certain pressure and temperature conditions. By controlling the pressure we can indirectly control the density of a fluid, its phase and in some cases its temperature. Pressure also allows us to control the safety of the process as it is potentially one of the most dangerous variables since an overpressure of a vessel or pipeline can cause leaks, or rupture and explosion of the vessel. PDVSA Technical Report (2010).

Absolute Pressure

Absolute or Barometric Pressure is the pressure measured taking into account the pressure of the earth's atmosphere; approximately 14.7 psi at sea level. It is denoted by the postfix abs or (absolute) after the unit of measure (psi abs).

Manometric or Relative Pressure

It is the pressure measured without taking into account the earth's atmospheric pressure. It is denoted by the postfix g (gauge) after the unit of measurement (psi g).

Flow rate

Flow is the amount of fluid passing through a section per unit time. It can be referred to as liquid or gas flow. It is usually identified with volumetric flow or volume passing through a given area in unit time. Less frequently, it is identified with the mass flow or mass passing through a given area in unit time.

The performance of some fluid meters is affected by the properties and conditions of the fluid; a basic consideration is whether the fluid is a liquid or a gas. Other factors that may be important are viscosity, temperature, corrosion, electrical conductivity, optical clarity, lubrication properties and homogeneity.

This factor determines the installation (pumps, pipelines, among others) required to transport liquids; it is the quantity that must be discharged. Among the flow units commonly used in the industry are: GPM gallons per minute, BPD barrels per day, CFD cubic feet per day, CFH cubic feet per hour, BPH barrels per hour, CFM cubic feet per minute, LbsH pounds per hour. PRLS Operating Manual (2010).

Gas Expansion
It consists of the sudden reduction of the pressure of a gas, usually in order to reduce its temperature. It is used in refrigeration systems and the refrigerant is subjected to this effect. GPSA (2004)

Heat transfer
When two bodies at different temperatures are placed in contact, the temperature of both tends to equilibrate. For the above process to occur there must be a transfer of heat from the hot body to the cold body. This amount of heat transfer is measurable and is usually expressed in BTU.

Heat flows, as a result of the temperature gradient, from the hot fluid to the cold fluid through a partition wall, which is called the heat transfer surface or area. Cengel (2009)

Heat transfer mechanisms

Heat transfer is an interaction between fluids or materials as a result of a temperature gradient between them. This interaction occurs through two different mechanisms: conduction and convection. PDVSA Process Design Manual (1995)

- Convection

The convection mechanism is strongly influenced by the flow pattern (fluid dynamics), but has an associated energy exchange from high to low temperature zones.

Convection is the transfer of heat from one point to another in a fluid, gas or liquid, due to the mixing and movement of the different parts of the fluid. There are two mechanisms of heat transfer by convection, called forced convection and natural convection. PDVSA Process Design Manual (1995)

- Driving

Conduction is essentially energy transfer by physical contact in the absence of movement of the material at the macroscopic level. This mechanism can occur in solids, liquids or gases. PDVSA Process Design Manual (1995)

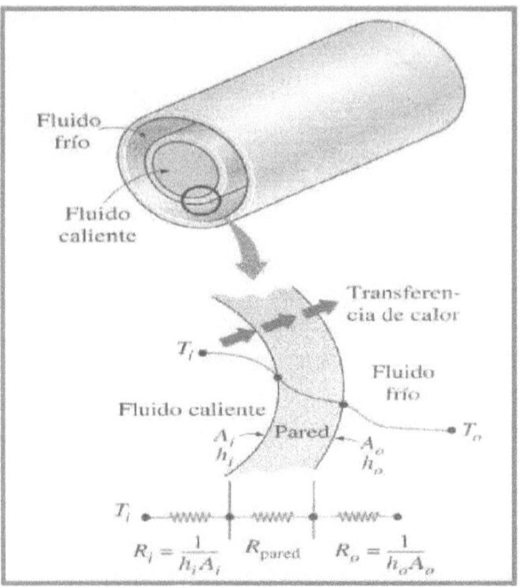

Figure 22 Heat transfer process. Source: Cengel 2009

As shown in Figure 6 the heat transfer from the hot fluid to the inner wall of the pipe is by convection and the following formula is used to determine this resistance:

$$R = \frac{1}{(h_i A_i)} \qquad \text{Ec.1}$$

The subscripts *i* and *o* refer to the inner and outer surfaces of the tube. Ai is the area or inner surface of the tube and is calculated: hi is the internal convection heat transfer coefficient or film coefficient, and is calculated:

$$A_i = \pi D_i L \qquad \text{Ec. 2}$$

K is the thermal conductivity of the fluid.
Nu is the number of nusselt
Whereas, through the inner wall of the pipe the heat transfer to the cold fluid is by conduction and this resistance is calculated by the following formula:

$$h_i = \frac{K}{D_i} Nu \qquad \text{Ec. 3}$$

$$R_{Pared} = \frac{Ln\,(D_o \,/\, D_i)}{2\pi K L} \qquad \text{Ec.4}$$

$$R_o = \frac{1}{(h_o A_o)} \qquad \text{Ec. 5}$$

K is the thermal conductivity of the wall material and L is the length of the pipe.

Finally, the heat transfer from the outer wall of the tube to the cold fluid is by convection, and this resistance is calculated by the following formula:

$$A_o = \pi D_o L \qquad \text{Ec. 6}$$

Where Ao is the external surface of the tube and is calculated as follows: **ho** is the external convection

$$h_o = \frac{K}{D_o} \, Nu \qquad \text{Ec. 7}$$

heat transfer coefficient or film coefficient.

Then the total thermal resistance is: (figure 1)

$$R_{total} = R_i + R_o + R_{Pared} \qquad \text{Ec.8}$$

Heat transfer process

In the previous section the different heat transfer mechanisms have been discussed, here we will discuss how heat transfer occurs through these mechanisms. There are two general types of processes; namely: Without phase change, also known as sensible heat and with phase change.

The non-phase change or sensible heat process, as its name suggests, involves fluid heating and cooling operations where the heat transfer results only in temperature changes; whereas in phase change, the operation results in a conversion from liquid to vapor or vapor to liquid; i.e., vaporization or condensation. Many applications involve both types of processes.

Sensitive heat

Most applications of non-phase change heat transfer processes involve the mechanism of forced convection transfer, both within tubes and on external surfaces. The convective heat transfer coefficient depends on fluid dynamics parameters, e.g. velocity.

Based on the fluid motion, the flow inside the pipes is divided into three flow regimes, which are measured by a dimensionless parameter, called Reynolds number, **which** is an indication of the turbulence of the flow.

Regarding the phenomenon of heat transfer by forced convection over external surfaces, it is important to mention that the heat transfer process is intimately related to the nature of the flow; for example, heat transfer over a tube bundle depends on the flow pattern and the degree of turbulence; i.e., it is a function of the 78 velocity of the fluid and the size and arrangement of the tubes. Because the complexity of the flow in these cases of heat transfer over external surfaces makes their analytical treatment difficult, the equations available for the calculation of the heat transfer coefficient have been developed entirely on the basis of experimental data.

Reynolds number

The Reynolds number is an indication of the type of fluid flow, laminar or turbulent. A large Reynolds number value indicates turbulent flow, while a small Reynolds number value indicates laminar flow.

When the Reynolds number is less than 2,100 the flow is laminar, when the Reynolds number is between 2,100 and 10,000 it is transitional flow, while it is turbulent flow if the Reynolds number is greater than 10,000.

For each of these flow regimes, semi-empirical equations have been developed which are used to adequately describe and predict the heat transfer in the region under consideration.

Reynolds Number for a Circular Pipe

The Reynolds number for a circular pipe is defined as:

$$Re = \frac{\rho\, V_m\, D}{\mu} \qquad\qquad \text{Ec.9}$$

Where **p** is the fluid density, **p** is the fluid viscosity and **D** is the hydraulic diameter or internal diameter and is calculated:

$$D_h = \frac{4A_c}{P} = \frac{(4\pi D^2/4)}{\pi D} \qquad\qquad \text{Ec.10}$$

Ac is the cross-sectional area of the pipe and **P** the perimeter **Vm** is the average velocity and its formula is given by:

$$Vm = \frac{\dot{V}}{At} \qquad\qquad \text{Ec.11}$$

Where **v** is the product flow rate **At** is the cross-sectional area, we obtain it as follows:

Nusselt number

This represents the enhancement of heat transfer through a fluid layer as a result of convection relative to conduction through the same layer. The higher the Nusselt number the more efficient the convective heat transfer. The calculation of Nu will depend on the Reynolds number calculation.

$$At = \frac{\pi}{4} \times D_i^2 \times N^o Tubos \qquad\qquad \text{Ec.12}$$

$$Nu = \frac{h\, L_c}{k} \qquad\qquad \text{Ec.13}$$

Where **Lc** is the characteristic length, **k** is the thermal conductivity of the fluid and *h* is the convective heat transfer coefficient.

Flow through pipe banks

The flow inside the tubes can be analyzed by considering the flow through only one of them and multiplying the results by the number of tubes, however this is not the case for the flow over the tubes, as they influence the flow pattern and the level of turbulence downstream and, consequently, the heat transfer to and from them.

In pipe banks the flow characteristics are denoted by the maximum velocity *Vmax* inside the bank rather than by the approximate velocity *V. Therefore,* the Reynolds number is defined on the basis of the maximum velocity as:

$$Re = \frac{\rho \, V_{max} \, D}{\mu} = \frac{V_{max} \, D}{v} \qquad \text{Ec. 14}$$

v is the kinematic viscosity In tube banks the Reynolds number is based on the maximum velocity *Vmax,* which is related to the approach velocity, *V*, according to:

Aligned Tubes:

$$V_{max} = \frac{S_t}{S_t - D} \, V \qquad \text{Ec.15}$$

Staggered with SD < (St + D) / 2

St is the transverse pitch and **SD** is the diagonal pitch.

$$V_{max} = \frac{S_t}{2(S_D - D)} \, V \qquad \text{Ec. 16}$$

The flow through banks of tubes is studied experimentally as it is too complex to be treated analytically. Mainly, the interest is in the average heat transfer coefficient for the entire bank of tubes, which depends on the number of rows along the flow as well as on the arrangement and size of the tubes. Several correlations have been proposed, all based on experimental data for the average Nusselt number for cross flow over tube banks. More recently, Zukauskas has proposed correlations whose general form is

$$Nu_D = \frac{hD}{k} = C \, Re_D \, P^n \, (Pr/Pr_s)^{0,25} \qquad \text{Ec.17}$$

$$T_m = \frac{T_i + T_e}{2} \qquad \text{Ec.18}$$

Where the values of the constants C, m and n depend on the value of the Reynolds number. Table 7-2 gives these correlations explicitly.

$$Nu_{D,\,NL} = FNu_D \qquad \text{Ec.19}$$

$$Q = h\,A_s\,\Delta T_{ln} \qquad \text{Ec. 20}$$

Table 2
Nusselt Number for Pipe Banks

Correlaciones del número de Nusselt para flujo cruzado sobre bancos de tubos, para $N > 16$ y $0.7 < Pr < 500$ (tomado de Zukauskas, Ref. 15, 1987)*		
Disposición	Rango de Re_D	Correlación
Alineados	0-100	$Nu_D = 0.9\ Re_D^{0.4}Pr^{0.36}(Pr/Pr_s)^{0.25}$
	100-1 000	$Nu_D = 0.52\ Re_D^{0.5}Pr^{0.36}(Pr/Pr_s)^{0.25}$
	1 000-2 × 10⁵	$Nu_D = 0.27\ Re_D^{0.63}Pr^{0.36}(Pr/Pr_s)^{0.25}$
	2 × 10⁵-2 × 10⁶	$Nu_D = 0.033\ Re_D^{0.8}Pr^{0.4}(Pr/Pr_s)^{0.25}$
Escalonados	0-500	$Nu_D = 1.04\ Re_D^{0.4}Pr^{0.36}(Pr/Pr_s)^{0.25}$
	500-1 000	$Nu_D = 0.71\ Re_D^{0.5}Pr^{0.36}(Pr/Pr_s)^{0.25}$
	1 000-2 × 10⁵	$Nu_D = 0.35(S_T/S_L)^{0.2}\ Re_D^{0.6}Pr^{0.36}(Pr/Pr_s)^{0.25}$
	2 × 10⁵-2 × 10⁶	$Nu_D = 0.031(S_T/S_L)^{0.2}\ Re_D^{0.8}Pr^{0.36}(Pr/Pr_s)^{0.25}$

Source: Data obtained from Cengel Heat Transfer.

All properties, except Pr, should be evaluated at the arithmetic mean of the fluid inlet and outlet temperatures determined from:

Where Ti and Te are the fluid temperatures at the inlet and outlet of the tube bank, respectively.
The average Nusselt number ratios in Table 2 are for tube banks with 16 or more rows. The average Nusselt number for tube banks with less than 16 rows is expressed as:

Where F is the correction factor.
Once the Nusselt number and thus the average heat transfer coefficient for the complete tube

bank is known, the rate of heat transfer can be determined from Newton's Law of cooling by means of an appropriate temperature difference, *AT* the rate of heat transfer into a tube bank or from it is determined from

Where *ATln* is the average logarithmic temperature difference h is the film coefficient and As is the heat exchange surface area.

Heat exchanger.

The generic name for a mechanical device, or equipment, designed to transfer heat between two or more fluid streams flowing through the equipment. PDVSA Process Design Manual (1995)

Exchangers are designed to meet specific requirements and are classified according to different criteria, such as heat transfer processes and mechanisms, degree of surface compactness, flow pattern, number of fluids, geometry and type of construction.

Operationalization of the variable

Table 2

General Objective	
To design a Pentane Refrigeration System in the Bajo Grande LPG Fractionation Plant.	
Conceptual Definition	**Operational definition**
The main objective is to cool the fractionated products in the plant for their subsequent storage in liquid state, thus reducing the size of the containers, the emission of vapors and optimizing the plant's production.	A refrigeration system consists of an organized network of equipment and accessories that make possible the thermal exchange between the product and the refrigerant, giving rise to the refrigeration cycle. For its design it is necessary to define the properties and conditions of the fluids involved in the process.

General Variable	Dimensions	Indicators
Pentane Cooling System	Operation	Refrigeration cycle. Refrigeration system Refrigerants used.

Operational and Design Parameters.	Operating and Design Pressure Operating and Design Temperature Operating and Design Temperature Physical and Thermodynamic Properties of Pentane Heat Transfer
Associated equipment and accessories	Heat Exchanger Throttle Valves Piping
Methods of Analysis	LMTD NTU

CHAPTER III
METHODOLOGICAL FRAMEWORK

Type Of Research

In the first place, it must be said that there is no single and exhaustive criterion that allows to cover the whole range of research that is carried out, they can be classified by simultaneously attending to the different aspects, however, there are discrepancies even among different authors, which makes the definition of the type of research to be carried out more ambiguous.

According to Hurtado (2010), "this type of research attempts to propose solutions to a given situation based on a previous research process. It starts from the identification of an event to be modified, and the descriptive diagnosis in which the research is initiated is made based on that event to be modified, in this sense such diagnosis is the one that allows corroborating that the proposal is really necessary".

Thus, taking into consideration the above, it can be said that the research is projective, because it consists of collecting the necessary data in order to be aware of the problem that manifests itself in terms of the lack of a pentane cooling system in the plant.

On the other hand, it has a descriptive aspect; according to Hurtado (2010), "it is that which aims to obtain a characterization of the event under study, to detail its qualities. It is not interested in studying chance relationships, in giving explanations or comparing groups; it only alludes to what something is like, how it manifests itself, or how it has been developing".

Research Design

The term "design" refers to the plan or strategy devised to answer the research questions. The design tells the researcher what he must do to achieve his study objectives and answer the questions he has set for himself; if the design is well conceived, the ultimate product of a study (its results) will have a greater chance of being valid. The design of this study is non-experimental, field and documentary. The design is a specific method, a series of successive and organized activities, which must be adapted to the particularities of each investigation.

The design of this research is non-experimental since, according to Tamayo and Tamayo (2007), it defines: "a non-experimental study is one where no situation is constructed, but where definitions of the system are observed, in order to try to adapt it to the situations required".

Likewise Hurtado (2010), "is one in which the researcher, despite wanting to verify hypotheses, does not have the possibility of manipulating the independent variables (explanatory processes), either because they have already occurred, because they are out of reach or for ethical reasons".

According to the above, this research project falls into this classification, since the variable under study is not manipulated, i.e., the refrigeration system is designed based on procedures already established in the plant and its parameters are defined without the use of an action and reaction process.

For its purpose, it is considered field research, as referred to by Hurtado (2010), "is that in which the researcher obtains data from direct sources in its natural context".

It is also considered field data because it is based on methods that allow the collection of information directly from the reality where it is presented, that is, the data were acquired directly from the field of operation.

Likewise, this study will be inherent with bibliographic and documentary material that will serve as a basis for the context of the research. This type of research is known as documentary research, which is defined by Hurtado (2010), "is that in which the researcher resorts to diverse documents as a source for the collection of data that will allow him to answer his research question".

In this sense, documentary research is based on bibliographic sources of all kinds, which support, by means of the information gathered, the objective(s) that have been set out in a given project.

For this reason, the research enters this phase, since the search, recovery and analysis of printed documents such as plans, manuals, guides, among others, were obtained at the Bajo Grande Fractionation Plant. As in all research, the purpose of this design is the contribution of new knowledge.

Reference population

According to Hurtado (2010), the reference population is the set of sources from which information will be obtained regarding the study population. In this sense, in this research, surveys and interviews were conducted with process engineers, technicians, plant operators, among others, who work at the ULÉ and Bajo Grande Fractionation plants, as well as at the La Salina Lake Terminal and the Western Cryogenic Complex Project, obtaining accurate information about the refrigeration systems.

Analysis Unit

It is a textual chain that can be differentiated from the rest of the document by referring to a particular subject. It is based on spatial and synthesis criteria (words, paragraphs, sections). It allows summarizing information and making it manageable (Hurtado, 2010).

According to the above, it is the totality of its elements of which can present certain characteristic susceptible to be studied, so it was used as a unit of analysis the design of a pentane refrigeration system in the Bajo Grande fractionation plant.

Data Collection Technique

The selection of data collection techniques and instruments involves determining by which means or procedures the researcher will obtain the necessary information to achieve the research objectives (Hurtado, 2010).

Tamayo and Tamayo (2007), regarding direct observation describe: "it is in which the researcher can observe and collect data through his own observation".

As can be inferred, the research is of this nature, since there is direct contact with the phenomenon or event that is being carried out, and in this way information is collected and recorded for subsequent analysis.

The starting point of the research was the elaboration of an extensive bibliographic compilation that incorporated and reviewed the theoretical and conceptual expertise developed in the field of research on the subject of the refrigeration system.

Likewise, it was necessary to collect information related to the operating conditions, design of the lines and equipment associated with the process, which included: Operations Manual of the Bajo Grande Fractionation Plant, equipment data and specifications, piping and instrumentation drawings, isometric drawings, degree theses and reports prepared by PDVSA personnel.

Additionally, all the processes related to the refrigeration system were studied in order to propose the most adequate design for the cooling of pentane in the plant. Likewise, the plant's available space was verified for the design of the necessary equipment and lines.

In addition, information was collected directly from the equipment, such as chillers, valves and piping located in the refrigeration plant area. It also included unstructured interviews with operations personnel and process engineers.

In the present work, information was collected from both primary and secondary sources.

Primary sources

It is the oral or written information that is collected directly by the researcher through accounts or writings transmitted by the participants in an event or occurrence. Tamayo (2008).

This type of source involves the use of techniques and procedures in which the researcher collects information directly either through direct observation in the field, interview with experts to share criteria, analysis and synthesis, the observation was carried out directly, in order to obtain information and in turn verify it, the interview was conducted as a process of unstructured verbal communication with specialized personnel in the area of fractionation processes.

Direct observation in the field

Tamayo (2008) defines direct observation as the use of the senses for the perception of facts or phenomena that surround or are of interest to the researcher, where for the present research the constant interaction of the researchers with the organization allowed them to directly perceive the information in the work area.

This technique was of great benefit because it allowed the researchers to learn more about the processes carried out at the Bajo Grande LPG fractionation plant.

Unstructured interview

It consists of freely formulating questions, based on the answers given by the respondent. There is no standardization of the form and the questions may vary from one respondent to another. Hurtado (2010).

Therefore the margin of freedom to formulate the questions under the application of this method and obtain the necessary answers, benefit the investigation as information is obtained concerning the refrigeration system and at the same time other questions arise based on the answers transmitted, and thus clarifying doubts that may arise in any part of the investigation.

Observation

According to Sabino (2005), he considers observation "as the process by which certain features existing in reality are perceived by means of a scheme and based on a certain purpose of the problem to be investigated".

The information collected was based on this technique since information was collected about all the elements that make up the refrigeration system of the plant, which serves as an operational reference for coupling this design to the existing system.

In this case, inspections and direct observations were made in order to carry out a visual evaluation of the plant's existing refrigeration system and to establish the possible options available.

Secondary Sources

It is written information that has been collected and transcribed by people who have received such information through other written sources, or by a participant in an event or occurrence. Hernández, Fernández and Baptista (2008). It represents the collection of information stored in written sources, which in this specific case, was extracted from texts or bibliography referring to the description of processes, manuals, standards and degree works.

Direct Observation Technique

Direct observation is defined as the use of the senses for the perception of facts or phenomena that surround or are of interest to the researcher (Tamayo, 2008). In this research, oral and written communication was carried out in which there was feedback between the sender and the receiver.

Investigation Procedure.

This research is based on a series of steps, where in the first place all the corresponding and necessary information is compiled for the elaboration of the first chapter, the problem statement. By means of the selected information, a clearer idea is obtained about the existing problem to which the solution is to be sought. In this way, the bases of the research begin to be elaborated, proposing general and specific objectives, which serve to delimit the same, and in this way to have an organized plan for the development of the research.

Secondly, an analysis for the fulfillment of the objectives is proposed, through which basic terms essential for the development of the objectives are defined.

On the other hand, research is taken into account as support material for the design, which serves as necessary input on the steps to be followed for its realization, and as information regarding the selection of different materials and components to be taken into account for the construction of the refrigeration system.

In the same way, a series of calculations will be made to guarantee the technical feasibility of each of the components selected for the complete system, to ensure its operation.

CHAPTER IV

ANALYSIS OF RESULTS

Description of the Bajo Grande plant refrigeration process (Area 300)

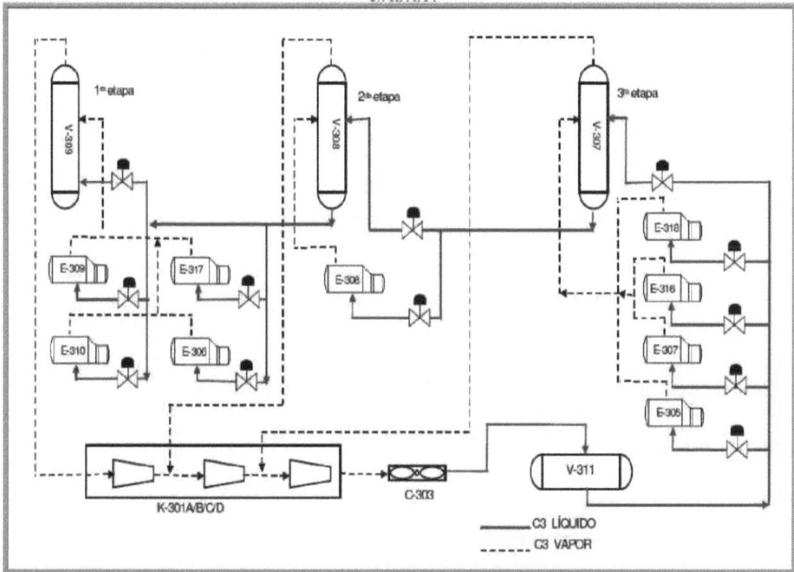

Figure 22. **Cooling system of the Bajo Grande fractionation plant (Area 300).** Source: PDVSA 2010

The refrigeration system at the Bajo Grande LPG Fractionation Plant is a closed system that uses propane as the refrigerant and a three-stage centrifugal compressor (K-301A/B/C/D). The function of this system is to cool the fractionated products in the plant prior to their storage in refrigerated tanks. This system has three refrigeration levels: 32 °C, 0 °C and -26 °C. The following table shows the properties of the propane refrigerant at the operating temperatures.

Table 1

Propane refrigerant at different levels of operation.

Level 1	Level 2	Level 3
32 °C (89.60 °F)	0°C *(32 °F)*	-26°C (-14.80 °F)
161.5 psig	60.98 psig	17.71 psig
-1165 BTU/lb	-1203 BTU/lb	-1230 BTU/lb

Source: Data obtained from document CCO-PIPC-03-PR-EST-10-0005 Rev. 0 PDVSA GAS (2010). Mendoza and Barreto (2011)

The propane leaving the compressor discharge K-301A/B/C/D is condensed in the air coolers C-

303 and then collected in the accumulator drum V-311; from there it is fed to the casings of the following coolers, located at the first temperature level: E-305 (propane), E-318 (n-butane), E-316 (i-Butane) and E-307 (gasoline). In each of the coolers, the propane vaporizes by receiving the heat from the products being cooled. The vapors resulting from the evaporation of propane from each cooler are collected in a common line and carried in the V-307 drum.

From the accumulator V-311, there is another liquid propane line which expands before reaching the economizer V-307. Its function is to guarantee the vapor flow to the third compression stage. From V-307 liquid propane flows to cooler E-308 (n-Butane cooler), located at the second temperature level. The vapor leaving the cooler goes to the economizer V-308. As in the first temperature level, liquid propane is extracted from V-307, which expands before reaching economizer V-308 to ensure vapor flow to the second compression stage.

The propane vapors generated in the E-308 cooler go to the V-308 economizer where they are mixed and constitute the suction of the second compression stage. The liquid propane from economizer V-308 feeds the third temperature level of the refrigeration system, which consists of the coolers: E-306 (propane), E-317 (i-Butane) and E-317 (i-Butane) and E-308 (propane).
309/E-310 (vapors from tank S-501). The vapors produced in the above-mentioned coolers go to the suction drum of the first stage of the V-309 compressor, thus closing the compression cycle and continuing to circulate through the process. To ensure vapor flow to the first compression stage, a liquid propane line is drawn and expanded before reaching V-309.

Operational parameters

Basis and premises.
• The pentane flow from the CCO will be 41711 Lb/hr added to 21451 Lb/hr of pentane from the ULÉ fractionation plant for a total of 63162 Lb/hr ("CCO-PIPC-03-PR-EST- 10-0005 Rev. 0").

Table 2
Flow of pentane from CCO and Bajo Grande

cco	ULE	Large Bass Case 1	LAMA	Cond	Large Bass Case 2	Gas Rico Note 2	Recruitmen t
LGN BEPD	LGN BEPD	LGN BEPD	LGN BEPD	LGN BEPD	LGN BEPD	LGN BEPD	C5 BEPD
35000	18000	5880	6000	3420	16180	72640	5096
Pentane Ib/hr	Swamp Ib/hr	Pentane ib/hr	Pentane Ib/hr	Pentane Ib/hr	Pentane Ib/hr	Pentane Ib/hr	Pentane Ib/hr
41711	21451	7005	8483	6520	19282	85705	46790

Source: Data obtained from CCO management Document ("CCO-PIPC-03-PR- EST-10-0005 Rev. 0").

- It is assumed that the pentane flow from the CCO will enter the Bajo Grande fractionation plant and will be temporarily stored in an accumulator tank, and then sent via level control to the new E-319 pentane cooler at the same pressure and temperature conditions as the pentane coming from the V-8331 depentanizer unit at 68.5 Psig and 120.5 °F respectively (See PFD "1150-01-70-P23-101 Rev.0)").

- The pentane outlet temperature of the E-319 chiller was set at 75 °F according to document "CCO-PIPC-03-PR-EST-10- 0005". And considering that the maximum pressure drop allowed for heat exchangers on the tube side is 5 Psi maximum according to PDVSA Standard MDP-05-E-01 Rev. 0, then the outlet pressure of the chiller is 63.5 Psig.

- The pentane flow rate coming from tower V-8331 (Bajo Grande Depentanizer Unit), assumed for the calculation is based on 1481 BEPD on the basis of what is established in the document "CCO-PIPC-03- PR-EST-09-0002 Rev. 0". This flow is equivalent to 13600 Lb/hr.

Table 3

Pentane flow from depentanizer unit V-8331

LGN de LAMA [BEPD]	2A - Carga Total a V-8331 [BEPD]	% Capacidad de diseño	6A - Pentanos en Destilado [BEPD]	% Reducción de Pentanos	13A - Gasolina de la V-8331 [BEPD]	% Reducción de Gasolina
0	3202	54%	817	55%	2384	22%
100	3228	55%	833	54%	2395	22%
200	3256	55%	849	53%	2407	21%
300	3284	56%	866	52%	2418	21%
400	3312	56%	883	51%	2429	21%
500	3340	57%	900	50%	2440	20%
600	3368	57%	917	49%	2452	20%
1100	3509	60%	1001	44%	2508	18%
1200	3537	60%	1018	44%	2519	18%
1500	3621	61%	1068	41%	2553	17%
1600	3649	62%	1085	40%	2564	16%
2100	3788	64%	1168	35%	2620	14%
2200	3816	65%	1185	34%	2631	14%
2500	3900	66%	1235	31%	2665	13%
2600	3927	67%	1251	31%	2676	13%
3100	4066	69%	1333	26%	2733	11%
3200	4094	70%	1350	25%	2744	10%
3500	4177	71%	1399	22%	2778	9%
3600	4205	71%	1416	21%	2789	9%
3800	4260	72%	1448	20%	2812	8%
3900	4287	73%	1464	19%	2823	8%
4000	4315	73%	1481	18%	2834	7%
4100	4343	74%	1496	17%	2846	7%
4200	4370	74%	1513	16%	2857	7%
4500	4453	76%	1562	13%	2891	6%
4600	4480	76%	1578	12%	2902	5%
5100	4617	78%	1658	8%	2959	3%
5200	4645	79%	1674	7%	2970	3%
5500	4727	80%	1723	4%	3004	2%

Source: Data obtained from document CCO-PIPC-03-PR-EST-09-0002 Rev. 0, from CCO management. Mendoza and Barreto (2011)

- The compositions of the pentane to be cooled in the large fractionation plant are specified in Table 1, and are supported in documents (400-CN-0001-01_0. Rev. 0) and (1150-01- 70-P09-CAL-001. Rev. 0).

Table 4

Composition of the pentane flow to be cooled at Bajo Grande

Component	Pentane CCO and ULÉ (Molar Frac.)	Pentane BG (Molar Fraction)
C1	0,0000	0,0000
C2	0,0000	0,0000
C3	0,0000	0,0000
IC4	0,0002	0,0002
N-C4	0,0056	0,0069
I-C5	0,4836	0,4961
N-C5	0,4582	0,4245
C6	0,0447	0,0720
C7	0,0066	0,0002
C8	0,0007	0,0000
C9	0,0002	0,0000
C10	0,0000	0,0000
TOTAL	0,9998	0,9999

Source: Data obtained from documents 400-CN-0001-01_0. Rev. 0 and 1150-01 70-P09-CAL-001. Rev. 0, at CCO Management. Mendoza and Barreto (2011)

- Since the highest percentage of substance corresponds to pentane (94%), the calculation of the properties of pentane will be made directly from the pressure-enthalpy diagram.

- The fluid inside the casing of the new E-319 pentane cooler will be propane since according to PDVSA MDP-05-E-01 Rev.0 standard the fluid to be vaporized must go through the casing while the fluid to be cooled goes through the tubes according to the same standard. Therefore, the cooling fluid is fixed through the casing and the pentane product through the tubes.

- The tube gauge of the new heat exchanger selected for the design will be BWG 14, with a nominal diameter of ¾ inches and 10 feet long and a tube pitch of 1" inch of carbon steel, 30° staggered triangular arrangement. This selection is based on TEMA (Tubular Exchanger Manufacturers Association) and PDVSA standard MDP-05-E-01 Rev. 0. (See Note 1 at the end of "operational parameters")

- The location of the new E-319 pentane chiller will be on the third refrigeration level due to the thermal availability of the chiller 2.24 MMBTU/hr. Therefore, the propane refrigerant inlet conditions to the new E-319 pentane chiller will be -26 °C (-14.80 °F) temperature and pressure of 17.71 Psig "CCO-PIPC-03-PR-EST-10-0005 Rev. 0".

Table 5

Thermal availability of the three cooling levels.

Cases	First Level	Second Level	Third Level
Case 1	4.68	0.66	12.89
Availability MMBTU/hr	4.39	1.57	3.40
Case 2	4.85	0.76	14.05
Availability MMBTU/hr	**4.22**	1.47	2.24

Source: Data obtained from documents CCO-PIPC-03-PR-EST-10-0005, at CCO Management. Mendoza and Barreto (2011)

•	The propane outlet temperature of the chiller will be simulated to find the temperature at which the propane vaporizes while maintaining constant pressure and thus ensure the flow of vapor to the compressors. A first approximation will be made based on the operating temperature of the suction drum of the first compression stage V-309 in charge of collecting the vapors of the third temperature level -10 °F. (see note 1).

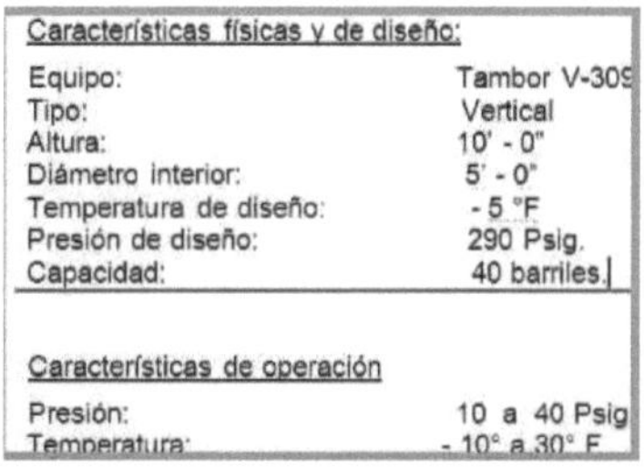

Características físicas y de diseño:
Equipo: Tambor V-309
Tipo: Vertical
Altura: 10' - 0"
Diámetro interior: 5' - 0"
Temperatura de diseño: - 5 °F
Presión de diseño: 290 Psig.
Capacidad: 40 barriles.
Características de operación
Presión: 10 a 40 Psig
Temperatura: - 10° a 30° F

Figure 23. **Operating characteristics of the V-309 economizer.** Source: Bajo Grande LPG fractionation plant operations manual "POPE 006_REV1". Mendoza and Barreto (2011)

•	The composition of the propane refrigerant for the refrigeration system of area 300 is supported in the procedures manual of the Bajo Grande fractionation plant document "POPE 006_REV1", and the document "Evaluation study of the existing refrigeration system of the Bajo Grande LPG plant to consider the installation of a new pentane chiller Product" CCO-PIPC-03-PR-EST-10-0005 shown below:

Table 6

Composition of Propane Refrigerant Area 300

<table>
<tr><td colspan="5" align="center"><u>Table 4. Composition of NGL and Refrigerant used.</u></td></tr>
<tr><td align="center">Components</td><td align="center">NGL Average
Note 2</td><td align="center">LAMA
Note 2</td><td align="center">Condensed
Note 2</td><td align="center">Propane
Refrigerant</td></tr>
<tr><td></td><td align="center">Molar Fraction</td><td align="center">Molar Frac</td><td align="center">Molar Frac</td><td align="center">Molar Frac</td></tr>
<tr><td align="center">Methane</td><td align="center">0.0000</td><td align="center">0.0000</td><td align="center">0.0000</td><td align="center">0.000</td></tr>
<tr><td align="center">Ethane</td><td align="center">0.0058</td><td align="center">0.0137</td><td align="center">0.0019</td><td align="center">0.033</td></tr>
<tr><td align="center">Propane</td><td align="center">0.5837</td><td align="center">0.4477</td><td align="center">0.1871</td><td align="center">0.953</td></tr>
<tr><td align="center">i-Butane</td><td align="center">0.0900</td><td align="center">0.1107</td><td align="center">0.0797</td><td align="center">0.012</td></tr>
<tr><td align="center">n-Butane</td><td align="center">0.1618</td><td align="center">0.2136</td><td align="center">0.1719</td><td align="center">0.002</td></tr>
<tr><td align="center">i-Pentane</td><td align="center">0.0540</td><td align="center">0.0661</td><td align="center">0.1000</td><td align="center">0.000</td></tr>
<tr><td align="center">n-Pentane</td><td align="center">0.0547</td><td align="center">0.0664</td><td align="center">0.1095</td><td align="center">0.000</td></tr>
<tr><td align="center">n-Hexane</td><td align="center">0.0329</td><td align="center">0.0818</td><td align="center">0.1564</td><td align="center">0.000</td></tr>
<tr><td align="center">n-Heptane</td><td align="center">0.0123</td><td align="center">0.0000</td><td align="center">0.1197</td><td align="center">0.000</td></tr>
<tr><td align="center">n-Octane</td><td align="center">0.0028</td><td align="center">0.0000</td><td align="center">0.0683</td><td align="center">0.000</td></tr>
<tr><td align="center">n-Nonane</td><td align="center">0.0016</td><td align="center">0.0000</td><td align="center">0.0040</td><td align="center">0.000</td></tr>
<tr><td align="center">n-Dean</td><td align="center">0.0003</td><td align="center">0.0000</td><td align="center">0.0015</td><td align="center">0.000</td></tr>
<tr><td align="center">n-C11</td><td align="center">0.0000</td><td align="center">0.0000</td><td align="center">0.0000</td><td align="center">0.000</td></tr>
</table>

Source: Data obtained from the Bajo Grande plant operations manual (POPE 006_REV1). Mendoza and Barreto (2011)

• Since the composition of the refrigerant propane is 95.3 %, the calculation of its properties as a pure substance in the pressure-enthalpy diagram is assumed.

• An estimate will be made to determine the length of the pipes associated to the new E-319 pentane cooler using Google Map software, the type of pipe will be Schedule 40 carbon steel because this type of pipes are commonly used within the company for this type of service. The diameter of the E-319 cooler nozzles obtained in the calculations will be taken as the same diameter.

• For the simulation and verification of the system, the Hysysys 7.2 simulator version 2010 will be used. The Pipephase 9.1 software will be used for pipe simulation and Autocad 2007 will be used for the construction of the process flow diagram (see note 2).

• The calculations for the sizing of the new E-319 pentane chiller are based on the Logarithmic Mean Temperature Difference (LMTD) method defined in Chapter II, since this method is applicable when the inlet and outlet temperatures of the fluids and the mass flow are known. Likewise, the recommendations of the PDVSA standard, also defined in said chapter, are taken into account for the calculation.

Note 1

Some of the above parameters are subject to modification as the calculations are developed, since they are based on the conditions available at the plant.

Note 2

The Hysys simulator will be used for the calculation of the system since they are calculated using rigorous methods that include equations of state and thermodynamic packages that are not related to the exchanger volumetry.

Calculations for the sizing of the new E-319 pentane chiller.

<u>Propane Data</u>

TC, ent (inlet temperature) = -14.80 °F

TC, salt (Outlet temperature) = -10 °F

K (Thermal Conductivity) = *0.068* BTU/ hr Foot °F

p (Dynamic Viscosity) = 1.14 x 10^{r4} Lb/ft sec

Cp (Specific Heat) = 0.3963 BTU/ Lb °F

p (Density) = 35.10 Lb / Foot^3

Pent (Chiller inlet pressure) = 17.71 Psig

Psal (Chiller outlet pressure) = 17.21 Psig

<u>Pentane data</u>

TC, ent (inlet temperature) = 120.5 °F

TC, salt (Outlet temperature) = 75 °F

K (Thermal Conductivity) = 0.06 BTU/ hr Pie °F

p (Dynamic Viscosity) = 1.2 x 10^{-4} Lb/ft sec

Cp (Specific Heat) = 0.57011 BTU/Lb °F

p (Density) = 37.03 Lb / Foot^3

Pent (Chiller inlet pressure)= 68.5 Psig

Psal (Chiller outlet pressure)= 63.5 Psig

m (Total mass flow of pentane) = 76762 Lb/hr = 8360 BPD

Calculation of the heat given up by pentane

First, the heat that the pentane must give up to cool to 75 °F is calculated. This equation does not assume a phase change and is therefore applicable for pentane flow.

$$\dot{Q} = \dot{m}_c \cdot C_{pc} \, (T_{c,\,Ent} - T_{c,\,Sal})$$

Where

mC is the Mass Flow of the hot fluid, **Tent** and **Tsal** are the inlet and outlet temperatures of the pentane. **Cpc** is the specific heat of the hot fluid, in this case, pentane.

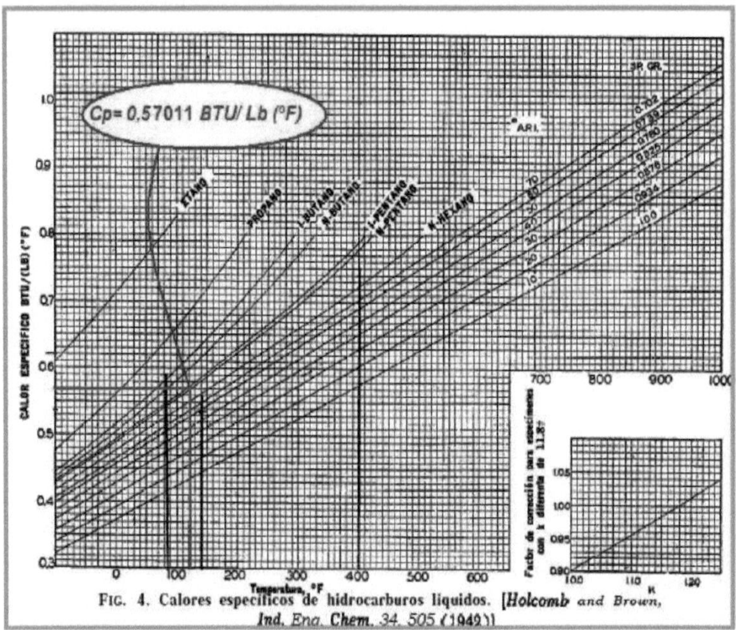

FIG. 4. Calores específicos de hidrocarburos líquidos. [Holcomb and Brown, Ind. Eng. Chem. 34, 505 (1942)]

Figure 24. **Calculation of the Cp of pentane at the chiller inlet @ 120.5 °F and 68.5 Psig.** Kern (1999).

When calculating the Cp, all the data is taken and substituted into the formula:

Data:

Cp = 0.57011 BTU/Lb °F

m = 76762 Lb/hr

Tent = 120.5 °F

Tsal = 75 °F

Q_p = 76762 Lb/hr x 0,57011 BTU/Lb °F x (120,5 – 75)°F

$$Q_p = 1.991.206{,}66 \text{ BTU/hr}$$

$$m_f\,(h_1 - h_2) = m_c\,(h_3 - h_4)$$

Calculation of the refrigerant mass flow.

The mass flow required to cool the pentane can then be calculated. This calculation is based on the first law of thermodynamics applied to a heat exchanger. Since propane is totally vaporized, a mass and energy balance must be performed in order to know the mass flow required to cool the product and to verify the heat flow between the products. The heat given up by the hot fluid (pentane) must be equal to the heat received by the cold fluid (propane). Cengel (2009)

Where **mf** and **mc** are the mass flow rates of the cold fluid (propane) and hot fluid (pentane) respectively. **h1** and **h2** are the inlet and outlet enthalpies of the propane heat exchanger while **h3** and **h4** are the inlet and outlet enthalpies of the pentane.

To find the enthalpy of each product, the pressure-enthalpy diagram of each product is used (Figure 1 and 5 Chapter II).

For the calculation of the propane enthalpy it is necessary to convert the pressures from Psig to Psia by adding 14.7 Psia, which is the atmospheric pressure measured at sea level. In addition, the exact temperature at which propane vaporizes was simulated in order to guarantee the vapor flow to the compressors, since the enthalpy pressure graph gives an approximate value and this parameter must be exact.

For the simulation we used as a first approximation the operating temperature of the economizer V-309 as established in the operational parameters -10 °F, since approximately at that temperature the vapors from the existing coolers are sent to that vessel.

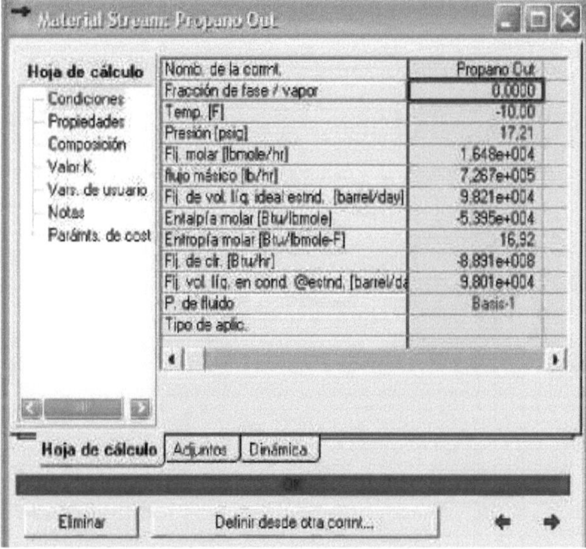

Figure 25 **Calculation of bubble temperature @ -10 °F and 17.21 Psig.** Hysys 7.2

As can be seen the bubble temperature is above the temperature taken as a reference. Therefore, a second approximation is made by specifying the vapor phase at 1 while keeping the pressure constant and in this way the simulator provides the exact temperature at which the propane is fully vaporized

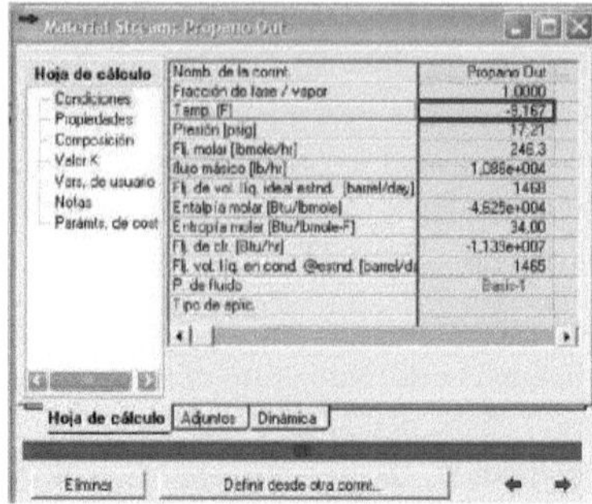

Figure 26 **Calculation of bubble temperature @ 17.21 Psig specifying a vapor fraction of 1 (total vaporization).** Hysys 7.2

It is observed that the vaporization temperature of propane is 9,167 °F for calculation purposes a temperature of -9 °F will be taken, which is close to the operating temperature of the suction drum V-309, this is theoretically feasible since the location of the new cooler E-319 does not provide greater variation with respect to the operating temperature of that vessel.

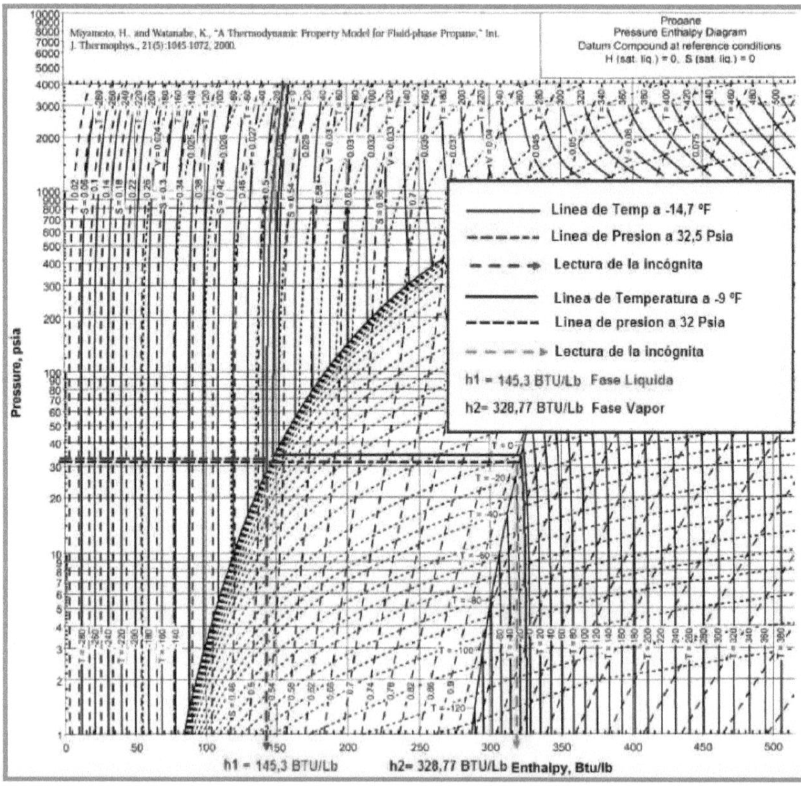

Figure 27. **Calculation of the phases and enthalpy of propane at the given conditions.** Source GPSA (2004)

Pent = 17.71 Psig + 14.7 = 32.41 ~ 32.5 Psia

Psal = 17.21 Psig + 14.7 = 31.91 ~ 32 Psia

h1 = 145.43 BTU/Lb

h2 = 328.77 BTU/Lb

For the calculation of the input and output enthalpy of pentane, the same consideration is made as for pentane.

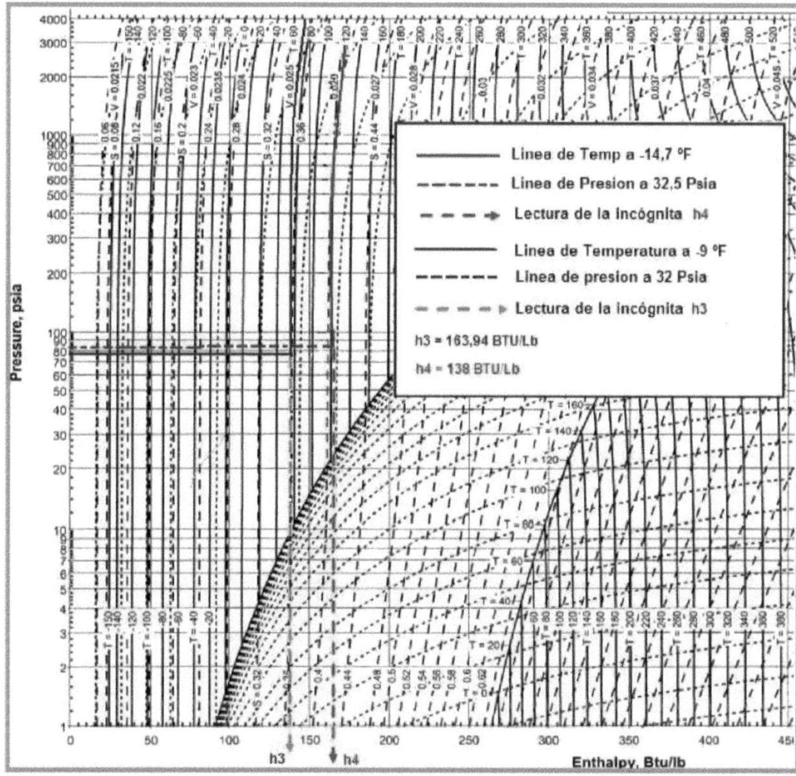

Figure 28. **Calculation of the phases and enthalpy of pentane at the given conditions.** Source GPSA (2004)

Pent = 68.5 Psig + 14.7 = 83.2 Psia

Psal = 63.5 Psig + 14.7 = 78.2 Psia

h3 = 163.94 BTU/Lb

h4 = 138 BTU/Lb

Once the inlet and outlet enthalpies are known, the following equation is applied and the mass flow rate is cleared as follows:

$$m_f = \frac{m_c\,(h3 - h4)}{(h2 - h1)}$$

Data

mc = 76762 Lb/hr

h3 = 163.94 BTU/Lb

h4 = 138 BTU/Lb

h1 = 145.43 BTU/Lb

h2 = 328.77 BTU/Lb

$$m_f = \frac{76762\ \text{Lb/hr} \times (163{,}94\ \text{BTU/Lb} - 138\ \text{BTU/Lb})}{(328{,}77\ \text{BTU/Lb} - 145{,}43\ \text{BTU/Lb})}$$

$$m_f = 10.860{,}73\ \text{Lb/hr}$$

The heat lost by the refrigerant under these parameters is found as follows:

$$Q = m_f \times (h_2 - h_1)$$

Data

mf = 10,860.73 Lb/hr *h1* = 145.43 BTU/Lb *h2* = 328.77 BTU/Lb

Qf = 10860,73 Lb/hr (328,77 BTU/Lb - 145,43 BTU/Lb)

$$Q_f = 1.991.206{,}24\ \text{BTU/hr}$$

The first law of thermodynamics can be demonstrated by this result, since, despite the fact that for pentane the enthalpy differential (Ah) is not taken but the specific heat Cp, it can be seen that there is a heat balance since **Qc = Qf**

Calculation of the logarithmic mean temperature difference (LMTD)
It will be used in the calculation of the area required for the design of the cooler.

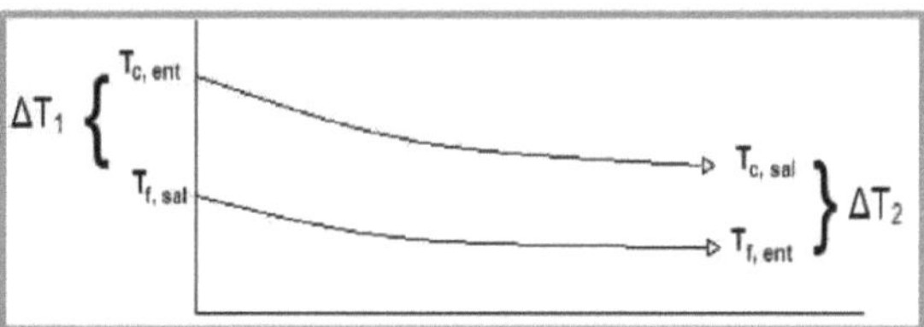

Figure 29. **Graphical analysis of the logarithmic mean temperature (LMTD) equation.** Source GPSA (2004)

Using the following equation, the logarithmic mean temperature difference is calculated:

$$\Delta T_{ml} = \frac{\Delta T_1 - \Delta T_2}{Ln(\Delta T_1 / \Delta T_2)}$$

Where ΔT_1 is the temperature difference between the hot fluid from inlet and the cold fluid outlet, i.e. $\Delta T_1 = T_{c,\,ent} - T_{f,\,sal}$ y ΔT_2 is the difference between the hot fluid at the outlet and the cold fluid at the inlet $\Delta T_2 = T_{c,\,sal} - T_{f,\,ent}$ (see figure 29)

Data

$$T_{c,\,ent} = 120,\,5\ ^{\circ}F$$
$$T_{f,\,sal} = -9\ ^{\circ}F$$
$$T_{c,\,sal} = 75\ ^{\circ}F$$
$$T_{f,\,ent} = -14,8\ ^{\circ}F$$

$$\Delta T_1 = T_{c,\,ent} - T_{f,\,sal} = \Delta T_1 = 120,5\ ^{\circ}F - (-\,9\ ^{\circ}F)$$
$$\Delta T_1 = 129,5\ ^{\circ}F$$
$$\Delta T_2 = T_{c,\,sal} - T_{f,\,ent} = \Delta T_2 = 75\ ^{\circ}F - (-14,8\ ^{\circ}F)$$
$$\Delta T_2 = 89,8\ ^{\circ}F$$

$$\Delta T_{ml} = \frac{129,5\ ^{\circ}F - 89,8\ ^{\circ}F}{Ln\left(\dfrac{129,5\ ^{\circ}F}{89,8\ ^{\circ}F}\right)}$$

$$\Delta T_{ml} = 108,44\ ^{\circ}F$$

As mentioned in the operational parameters, the heat exchanger has two passes through the tubes and one through the casing, for this reason the flow inside the chiller is considered as cross flow, for this type of heat exchanger the logarithmic mean temperature must be corrected by multiplying it by **F** which is the correction factor and is found by means of the following graph:

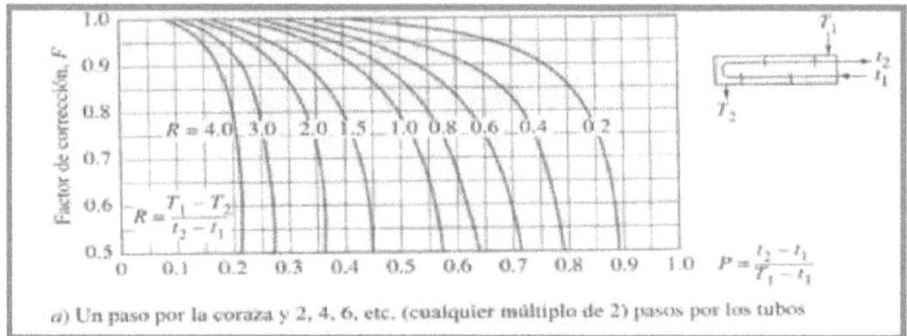

Figure 30. Graph for calculating the correction factor F. Source GPSA (2004)

$$P = \frac{t_2 - t_1}{T_1 - t_1} \qquad\qquad R = \frac{T_1 - T_2}{t_2 - t_1}$$

Where **P** and **R** are the ratios between two temperatures (i.e. it relates the temperatures of the casing and the tubes).

T is the temperature of the casing and **t** is the temperature of the tubes, sub indexes 1 and 2 indicate the inlet and outlet respectively.

Data
$t_1 = 120, 5\ ^\circ F$
$t_2 = 75\ ^\circ F$
$T_1 = -14,8\ ^\circ F$
$T_2 = -9\ ^\circ F$

$$R = \frac{-14,8\ ^\circ F - (-9\ ^\circ F)}{75\ ^\circ F - 120, 5\ ^\circ F} \qquad\qquad P = \frac{75\ ^\circ F - 120, 5\ ^\circ F}{-14,8\ ^\circ F - (-9\ ^\circ F)}$$

$$P = 0,51 \ y \ R = 0,13$$

When entering the table and reading the results, the value of **F** is approximately 0.996.

Calculation of heat exchanger area

Next, the exchange area necessary to size the new E-319 pentane cooler is found. For this calculation it is necessary to assume the value of the global heat transfer coefficient U, existing in tables, and in this way obtain the real U and substitute it again in the formula to find the real heat transfer area, as suggested in the PDVSA MDP-05-E-01 Rev. 0 standard described in chapter 2.

Kern (1999) classifies the value of the heat transfer coefficients U for light, medium and heavy organic substances; light organic substances are those with viscosities up to 0.5 cP and the design U is between 40 and 75 BTU/hr ft^2 °F, medium organic substances have viscosities between 0.5 cP and 1 cP and the U for these substances is between 20 and 60 BTU/hr ft °F, while heavy organic substances have viscosities greater than 1 cP and the design U for these substances is between 20 and 60 BTU/hr ft °F.

between 0.5 cP and 1 cP and the U for these substances is between 20 and 60 BTU/hr ft^2 °F , while heavy organic substances have viscosities greater than 1 centipoise (cP) and the overall transfer coefficient (design) is between 10 and 40 BTU/hr ft^2 °F (see Figure 31).

TABLA 8. VALORES APROXIMADOS DE LOS COEFICIENTES TOTALES **PARA DISEÑO.** MS **VALORES** INCLUYEN UN FACTOR DE OBSTRUCCION TOTAL DE 0.003 Y **CAIDA** DE PRESION PERMISIBLE DE 5 A 10 **LB/PLG**^{2} EN LA CORRIENTE QUE CONTROLE

Intercambiadores

Fluido caliente	Fluido frío	U_D total
Agua	Agua	2S0-500 *
Soluciones acuosas	Soluciones acuosas	2So-500 *
Sustancias orgánicas ligeras	Sustancias orgánicas ligeras	40-75
Sustancias orgánicas medias	Sustancias orgánicas medias	20-60
Sustancias orgánicas pesadas	Sustancias orgánicas pesadas\	**10-40**
Sustancias orgánicas pesadas	Sustancias **orgánicas** ligeras [	30-60
Sustancias orgánicas ligeras	Sustancias orgánicas pesadas/	10-40

[1] Las **sustancias orgánicas** ligeras son fluidos con viscosidades menores de 0.5 **centipoises** e **incluyen** benceno, tolueno, acetona, etanol, **metil-etil-cetona,** gasolina, **kerosén** y nafta.

[2] *Lar* sustancias **orgánicas** medias **tienen** viscosidades de 0.5 a 1.0 **centipois** e incluyen **kerosén, strawoil, gasoil** caliente, aceite de **absorbedor** caliente **y** algunos crudos.

[3] **Sustancias orgánicas pesadas** tienen viscosidades mayores de 1.0 **centipois** e incluyen **gasoil frío,** aceites lubricantes, **petróleo** combustible, **petróleo crudo** reducido, breas **y** asfaltos.

[4] Factor de **obstrucción** 0.001.

[5] **Caída** de presión de 20 a 30 **lb/plg**^{2}.

[6] Estas tasas **están** influenciadas grandemente por la presión de **operación.**

Figure 31. **Approximate values of total coefficients for design.** Source Kern (1999)

Both propane and pentane are light organic substances because their viscosities are in that range, therefore a first approximation is assumed at 40 BTU/hr ft^2 °F and the heat transfer area is calculated from the following equation and **As** is cleared.

Note 4: Some of the results generated in the calculations developed below may vary due to the assumption of the global transfer coefficient U, since they must be recalculated to find the real exchange area, which will allow sizing the new E-319 pentane cooler.

Where **As** is the heat transfer area, **U** is the total heat transfer coefficient.

Data

Calculation of the number of tubes (see note 4)

When the heat transfer area is obtained, the number of tubes can be found by the equation:

Where **P** is the perimeter $P = \pi D_o,$ **L** is the length of the pipe. By subtracting **No.**

The formula is as follows:

$$Q = U \times As \times \Delta T_{ml} \times F$$

Data

As = 461.37 ft^2

Q = 1.991.206,66 BTU/hr

ΔT_{ml} = 108,44 °F

F = 0,996

U = 40 BTU/hr Pie^2 °F

As = 461,37 pie^2

$$As = \frac{1.991.206,66 \ BTU/hr}{40 \ BTU/hr \ Pie^2 \ °F \times 108,44 \ °F \times 0,996}$$

$$\boxed{As = 461,37 \ pie^2}$$

$$\boxed{As = P \times L \times N° \ de \ tubos \times N° \ de \ pasos}$$

$$\boxed{N° \ de \ tubos = \frac{As}{N° \ de \ pasos \times \pi D_o \times L}}$$

N° de pasos = 2

D_o = 0,0625 pies^2

π = 3,14

L = 10 pies

$$N° \ de \ tubos = \frac{461,37 \ pie^2}{2 \times 3,14 \times 0,0625 \ pie^2 \times 10 \ pies}$$

$$\boxed{N° \ de \ tubos = 117,39 \sim 117 \ tubos. \ (Nota \ 4)}$$

In the operational parameters, it was proposed that the exchanger has 2 passes through the tubes and one through the casing, i.e., the resulting number of tubes is 117 tubes going and 117 tubes returning.

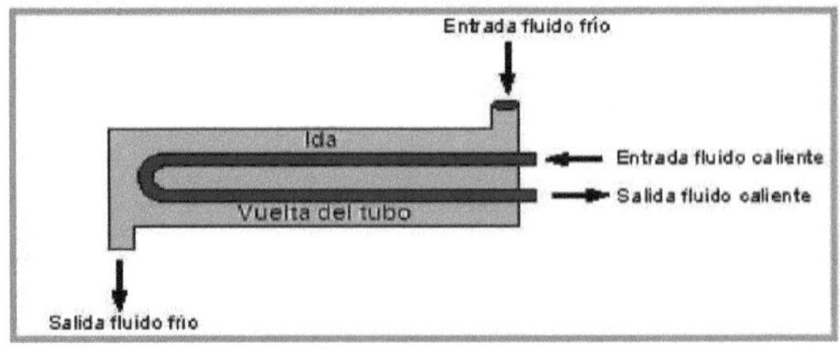

Figure 32. **Heat exchanger with two passes through the tubes and one pass through the shell.**
Source Cengel (2009)

Figure 33. **Heat exchanger with two passes through the tubes and one pass through the casing.** Source
PDVSA Gas (2010)

Calculation of the overall heat transfer coefficient U

$$U = \cfrac{1}{\left(\dfrac{1}{hi} \times \dfrac{Ao}{Ai}\right) + \left(Rfi \times \dfrac{Ao}{Ai}\right) + Rw + \dfrac{1}{ho} + Rfo}$$

To calculate U it is necessary to find the sum of all the resistances according to the data available,
first the wall resistance **Rw** is found.

$$R_{Pared} = \frac{Ln\,(D_o / D_i)}{2\pi KL}$$

Data

Do = 0.0625 ft

Di = 0.049 ft

K = 26 BTU/hr ft °F

L = 10 ft

Calculation of the internal area of the tubes

In chapter II the equation to find the area of the tubes was presented, however, it only corresponds
to the area of a single tube of a single pass.

$$R_{Pared} = \frac{Ln\left(\dfrac{0{,}0625 \text{ pies}}{0{,}049 \text{ pies}}\right)}{2\pi \times 26 \text{ BTU/hr pies °F} \times 10 \text{ pies}}$$

$$Rw = 1{,}49 \times 10^{-4}$$

Therefore, since the number of tubes and the number of steps are defined, this equation must be multiplied by the number of tubes and the number of steps.

$$A_I = \pi D_i \times L \times N^\circ\ tubos \times N^\circ\ de\ pasos$$

Data

$\pi = 3,14$

$Di = 0.049\ ft$

$L = 10\ ft$

№ **tubes** = 117

No. of steps = 2

Ai = 3.14 x 0.049 ft x *10 ft x* 117 x 2

$$A_i = 360{,}03\ pies$$

Calculation of the internal film coefficient "hi".

$$h_i = \frac{K}{D_i}\ Nu$$

The calculation of the film coefficient will depend directly on the Nusselt number, and in turn

$$Re = \frac{\rho\ V_m\ D}{\mu}$$

Nusselt will depend on the Reynolds number.

Where **Re** is the Reynolds number ρ is the density of the fluid **D** is the internal diameter of the tube μ the dynamic or absolute viscosity of the fluid passing through the tubes.

The dynamic viscosity is found by means of Figure 4 in the second chapter.

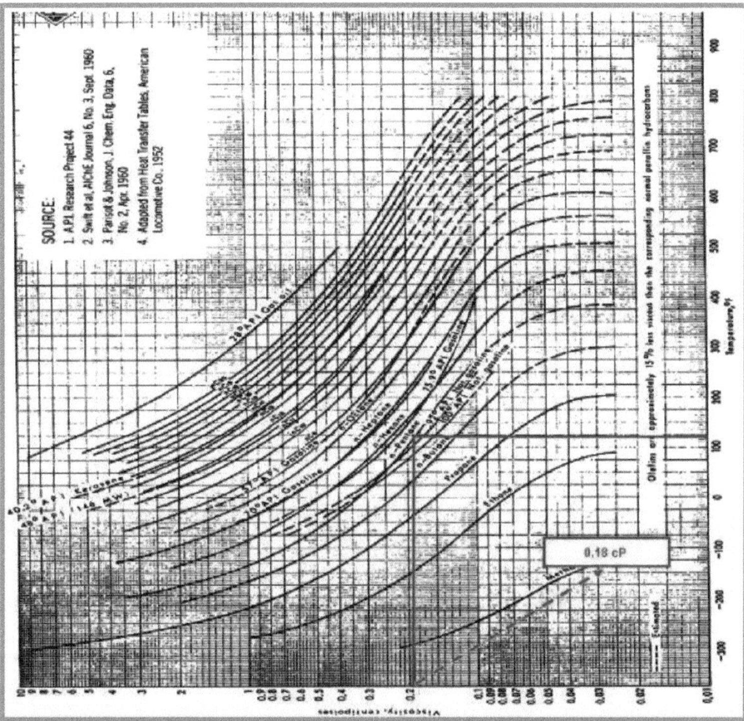

Figure 34. **Calculation of the dynamic viscosity of pentane.** Source GPSA (2004)

The viscosity obtained in the graph is approximately 0.18 cP. Next, the density is found, for the calculation of this unknown, the pressure-enthalpy diagram of the GPSA is applied in which the specific volume of the fluid is found at the pressure and temperature conditions of 68.5 Psig and 120.5 °F respectively, since the density is the inverse of the specific volume (1/v).

P = 68.5 Psig + 14.7 = 83.2 Psia

T = 120.5 °F

Figure 35. **Calculation of the specific volume using the pressure-enthalpy diagram.** Source GPSA (2004)

The specific volume is 0.027 Feet³ /Lb. by applying formula 1 given in chapter II, the density can be found.

$$\rho = 1/v$$

ρ = 1 / 0,027 Pies³/Lb

Then the average velocity **Vm** of the fluid is found.

The cross-sectional area of the tubes, i.e. the area through which the fluid passes, must first be known.

Data $\boxed{\rho = 37{,}03\ Lb/pie^3}$

Data $\boxed{Vm = \dfrac{\dot{V}}{At}}$

= 8360 BPD = 0.5433 feet /sec^3

At = 0.22 ft^2

= 8360 BPD = 0.5433 ft /sec^3

$$\boxed{At = \frac{\pi}{4} \times D_i^2 \times N^o Tubos}$$

Data

$_{Di}$= 0.049 ft

N°tubes = 117

p = *37.03 Lb/ft^3*

μ = 0.18 cP = 1.2 x 10^{-4} Lb/ft sec

At = 0.22 ft$^{2\backslash}$

$$At = \frac{\pi}{4} \times (0{,}049\ pies)^2 \times 117$$

$$\boxed{At = 0{,}22\ pies^2}$$

By knowing the density ρ of the pentane, the dynamic viscosity μ and the average fluid velocity, Reynolds can be determined.

Data

p = *37.03 Lb/ft^3*

μ= 0.18 cP = 1.2 x 10^{-4} Lb/ft sec

Di = 0.049 ft

Vm = 2.47 ft/sec

$$Vm = \frac{0{,}5433\ pies^3/seg}{0{,}22\ pies^2}$$

$$\boxed{Vm = 2{,}47\ pie/seg}$$

$$Re = \frac{37{,}03 \; Lb/pie^3 \times \; 2{,}47 \; pie/seg \times 0{,}049 \; pies}{1{,}2 \times 10^{-4} \; Lb/ \; pié \; seg}$$

$$\boxed{Re = 37.333{,}33}$$

As can be seen the Reynolds number is greater than 10,000 then we are in the presence of turbulent flow, for this type of regime the following relationship is used to find the Nusselt number **Nu**:

$$\boxed{Nu = 0{,}023 \times Re^{0,8} \times Pr^{n}}$$

Where **Pr** is the prandtl number and **n** = 0.4 for the heating of the fluid and 0.3 for the cooling of the fluid inside the tubes. In this case the fluid passing through the tubes (pentane) is cooled, i.e. the ratio to be used is 0.3. The prandtl number is calculated to know the nusselt number. To apply the prandtl equation it is necessary to find the specific heat and thermal conductivity, for both unknowns are used graphs 2 and 3 exposed in Chapter II.

$$\boxed{Pr = \frac{\mu \, Cp}{k}}$$

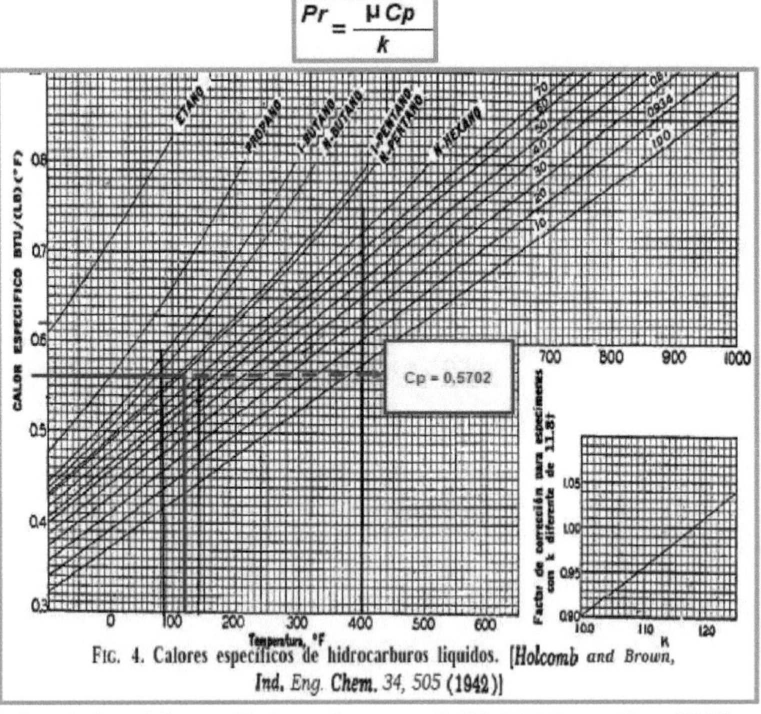

Fig. 4. Calores específicos de hidrocarburos líquidos. [Holcomb and Brown, Ind. Eng. Chem. 34, 505 (1942)]

Figure 36. **Calculation of the specific heat of pentane @ 120.5 °F.** Source Kern (1999)

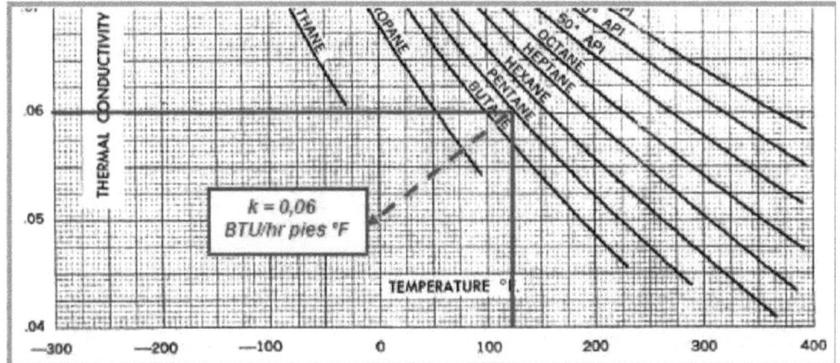

Figure 37. **Calculation of thermal conductivity of pentane @ 120.5 °F.** Source: THEME (1999)

Data

Cp = 0.5702 BTU/Lb

μ = 1.2 x 10^{-4} = 0.432 lb/ft hr

k = 0.06 BTU/hr ft °F

$$Pr = \frac{0,432 \text{ lb/pie hr} \times 0,5702 \text{ BTU/Lb}}{0,06 \text{ BTU/hr pies °F}}$$

$$\boxed{Pr = 4,1}$$

By finding the prandtl number, the nusselt number is found, which will be the final step to know the value of the internal film coefficient **hi**

$$Nu = 0,023 \times 37.333,33^{0,8} \times 4,1^{0,3}$$

$$\boxed{Nu = 159,67}$$

Data

k = 0.06 BTU/hr ft °F

Di = 0.049 ft

Nu = 159.67

$$h_i = \frac{0,06 \text{ BTU/hr pies °F}}{0,049 \text{ pies}} \times 159,67$$

$$\boxed{hi = 195,44 \text{ BTU/Lb pie}^2 \text{ °F}}$$

Calculation of the external film coefficient "*ho*".

To find **ho the calculation of** the Reynolds number, prandtl and nusselt is performed as was done to find the internal film coefficient **hi**. It is taken into account that the refrigerant propane moves through a bank of tubes, therefore, it must be considered the behavior that has the fluid through this bank of tubes. For this purpose, the arrangement of the tubes and their disposition are analyzed. In the operational parameters it was defined that the arrangement of the tubes of the new E-319 pentane cooler is triangular of 30° with staggered tubular arrangement. For this reason, the following consideration is made (see figure 36):

The PDVSA standard suggests a 30° triangular arrangement because square arrangements require a larger volume for the casing and triangular arrangements allow for higher heat transfer coefficients compared to square arrangements.

The longitudinal pitch **SL** is equal to the transverse pitch **ST** since the center-to-center spacing of each tube is 1", therefore **SL = ST** and its value is 1".

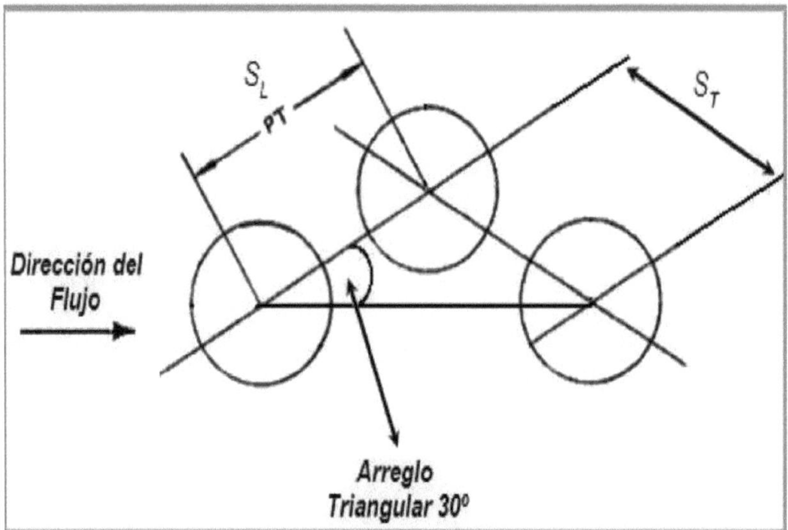

Figure 38. **Arrangement of the new E-319 pentane cooler.** Source: Mendoza and Barreto (2011)

$$Re = \frac{\rho \times Vmax \times Do}{\mu}$$

The velocity (Vmax) of liquid fluids according to the PDVSA standard is between 5 and 15 ft/sec of which the lowest 5 ft/sec is chosen.

The pressure-enthalpy diagram is used to find the density **p** of propane at the inlet conditions ***Pent*** = 17.71 Psig + 14.7 = 32.41 ~ 32.5 Psia and **Tent** = -14.8 °F.

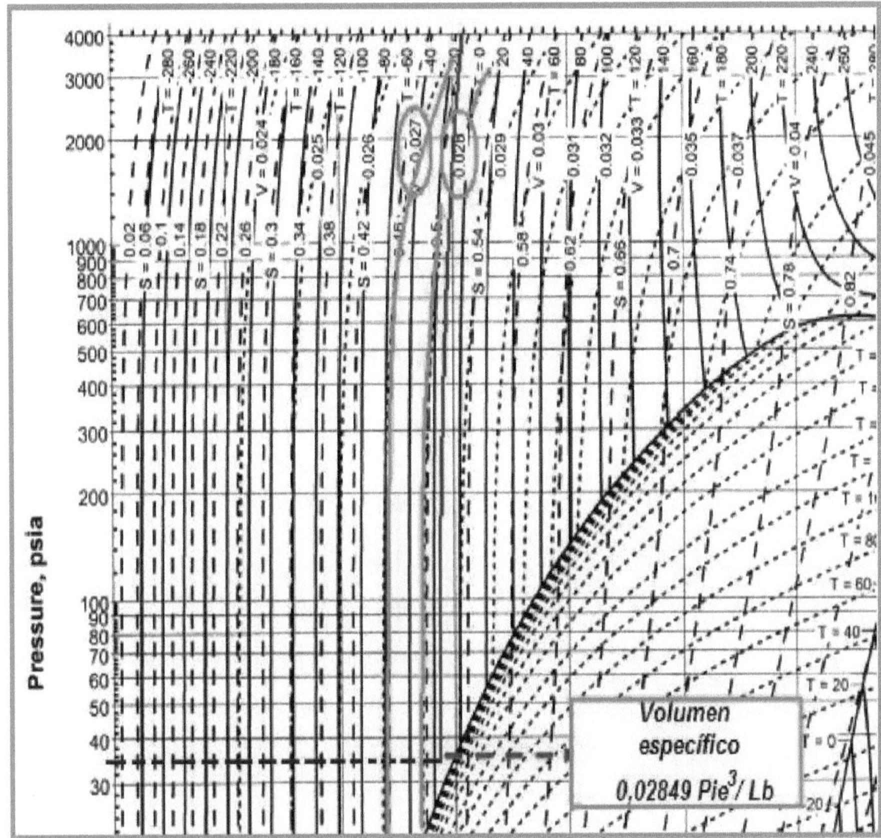

Figure 39.**Calculation of the specific volume of propane @-14.7 °F and 32.5 Psia.**
Source: Mendoza and Barreto (2011)

$$\rho = \frac{1}{0{,}02049 \ \text{Pie}^3 \ / \ \text{Lb}}$$

$$\rho = 35{,}10 \ \text{Lb/pie}^3$$

Next, the dynamic or absolute viscosity μ of propane is found.

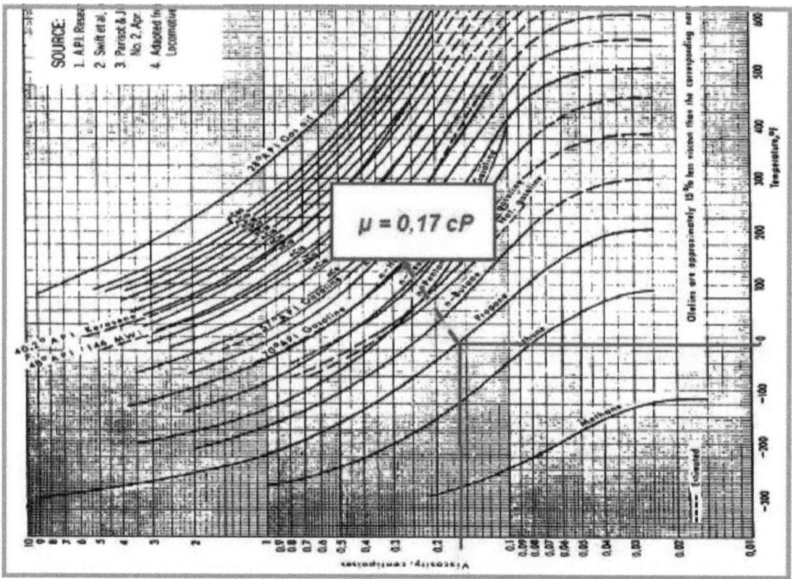

Figure 40. **Calculation of dynamic viscosity.** Source GPSA (2004)

Data

Vmax = 5 ft / sec

μ = 0.17 cP = 1.14 x 10^{-4} Lb/ft sec

ρ = 35.10 Lb/ft^3

Do = 0.0625 ft

$$Re = \frac{35{,}10 \ \text{Lb/pie}^3 \ x \ \ 5 \ \text{pie / seg} \ x \ 0{,}0625 \ \text{pies}}{1{,}14 \ x \ 10^{-4} \ \text{Lb/pie seg}}$$

$$Re = 96.218{,}07$$

With this result it is deduced that the flow inside the pipe bank is turbulent since Reynolds > 10000.

The number of nusselts for staggered pipe banks is between 1000 and 20000 as can be seen in the following table described in chapter II.

Correlaciones del número de Nusselt para flujo cruzado sobre bancos de tubos, para N > 16 y 0.7 < Pr < 500 (tomado de Zukauskas, Ref. 15, 1987)*		
Disposición	Rango de Re_D	Correlación
Alineados	0-100	$Nu_D = 0.9 \, Re_D^{0.4} Pr^{0.36} (Pr/Pr_s)^{0.25}$
	100-1 000	$Nu_D = 0.52 \, Re_D^{0.5} Pr^{0.36} (Pr/Pr_s)^{0.25}$
	1 000-2 × 10⁵	$Nu_D = 0.27 \, Re_D^{0.63} Pr^{0.36} (Pr/Pr_s)^{0.25}$
	2 × 10⁵-2 × 10⁶	$Nu_D = 0.033 \, Re_D^{0.8} Pr^{0.4} (Pr/Pr_s)^{0.25}$
Escalonados	0-500	$Nu_D = 1.04 \, Re_D^{0.4} Pr^{0.36} (Pr/Pr_s)^{0.25}$
	500-1 000	$Nu_D = 0.71 \, Re_D^{0.5} Pr^{0.36} (Pr/Pr_s)^{0.25}$
	1 000-2 × 10⁵	$Nu_D = 0.35(S_T/S_L)^{0.2} Re_D^{0.6} Pr^{0.36} (Pr/Pr_s)^{0.25}$
	2 × 10⁵-2 × 10⁶	$Nu_D = 0.031(S_T/S_L)^{0.2} Re_D^{0.8} Pr^{0.36} (Pr/Pr_s)^{0.25}$

Figure 41. **Nusselt number correlations for cross flow over pipe banks.** Source: Cengel (2009)

$$Nu = 0{,}35 \, (S_T/S_L)^{0.2} \times Re^{0.6} \times Pr^{0.36} \times (Pr/Pr_s)$$

To develop this equation it is necessary to find the prandtl number **Pr** which depends on the arithmetic mean temperature of the fluid determined from:

$$T_m = \frac{T_i + T_e}{2}$$

Where **Ti** and **Te** are the temperatures at the inlet and outlet of the tube bank, respectively.

Data

Ti = -14.8 °F

Te= -9 °F

$$T_m = -11{,}9 \text{ °F}$$

Then

TABLA A-12I

Propiedades del propano saturado

Temp. °F	Presión de saturación P, psia	Densidad ρ, lbm/ft³		Entalpía de vaporización h_g, Btu/lbm	Calor específico c_p, Btu/lbm·°F		Conductividad térmica k, Btu/h·ft·°F		Viscosidad dinámica μ, lbm/ft·h		Número de Prandtl Pr		Coeficiente de expansión volumétrica β, 1/R	Tensión superficial, lbf/ft
		Líquido	Vapor		Líquido	Vapor	Líquido	Vapor	Líquido	Vapor	Líquido	Vapor	Líquido	
−200	0.0201	42.06	0.0003	217.7	0.4750	0.2595	0.1073	0.00313	1.8042	0.01004	1.991	0.833	0.00083	0.001890
−180	0.0752	41.36	0.0011	213.4	0.4793	0.2680	0.1033	0.00347	1.4188	0.01071	6.582	0.826	0.00086	0.001780
−160	0.2307	40.65	0.0032	209.1	0.4845	0.2769	0.0992	0.00384	1.1518	0.01139	5.626	0.821	0.00088	0.001671
−140	0.6037	39.93	0.0078	204.8	0.4907	0.2866	0.0949	0.00423	0.9576	0.01209	4.951	0.818	0.00091	0.001563
−120	1.389	39.20	0.0170	200.5	0.4982	0.2971	0.0906	0.00465	0.8106	0.01280	4.457	0.817	0.00094	0.001455
−100	2.878	38.46	0.0334	196.1	0.5069	0.3087	0.0863	0.00511	0.6961	0.01352	4.087	0.817	0.00097	0.001349
−90	4.006	38.08	0.0453	193.9	0.5117	0.3150	0.0842	0.00534	0.6477	0.01389	3.936	0.819	0.00099	0.001297
−80	5.467	37.70	0.0605	191.6	0.5169	0.3215	0.0821	0.00559	0.6042	0.01426	3.803	0.820	0.00101	0.001244
−70	7.327	37.32	0.0793	189.3	0.5224	0.3284	0.0800	0.00585	0.5648	0.01464	3.686	0.822	0.00104	0.001192
−60	9.657	36.93	0.1024	186.9	0.5283	0.3357	0.0780	0.00611	0.5289	0.01502	3.582	0.825	0.00106	0.001140
−50	12.54	36.54	0.1305	184.4	0.5345	0.3433	0.0760	0.00639	0.4961	0.01540	3.490	0.828	0.00109	0.001089
−40	16.05	36.13	0.1641	181.9	0.5392	0.3513	0.0740	0.00668	0.4660	0.01579	3.395	0.831	0.00112	0.001038
−30	20.29	35.73	0.2041	179.3	0.5460	0.3596	0.0721	0.00697	0.4382	0.01619	3.320	0.835	0.00116	0.000987
−20	25.34	35.31	0.2512	176.6	0.5531	0.3684	0.0702	0.00728	0.4126	0.01660	3.253	0.840	0.00119	0.000937
−10	31.3	34.89	0.3063	173.8	0.5607	0.3776	0.0683	0.00761	0.3887	0.01701	3.192	0.845	0.00123	0.000887
0	38.28	34.46	0.3703	170.9	0.5689	0.3871	0.0665	0.00794	0.3666	0.01743	3.137	0.850	0.00127	0.000838

Figure 42. Prandtl numbers for saturated propane in liquid and vapor phase. Source: Cengel (2009)

Prandtl	Temperature
Y1 3,192	X1 -10 °F
Y2 ?	X2 -11.9 °F
Y3 3,253	X3 -20 °F

$$Y_2 = 3{,}204 = Pr$$

Prs must be evaluated at **Ts** i.e. -14.8 °F, therefore it is interpolated again to find its value.

Prandtl	Temperature
Y1 3,192	X1 -10 °F
Y2 ?	X2 -14.8 °F
Y3 3,253	X3 -20 °F

$$Pr_s = 3{,}221$$

With all the data, the values are substituted into the equation

$$Nu = 0{,}35\ (ST/SL)^{0,2} \times Re^{0,6} \times Pr^{0,36} \times (Pr/Prs)$$

Data

St = 0.083 ft

SL = 0.083 ft

Re = 96,218.07
Prs = 3,221
Pr = 3.204

$$\boxed{Nu = 518,79}$$

Finally, the **ho** film coefficient can be found.

$$\boxed{h_o = \dfrac{K}{D_o}\ Nu}$$

Data

k = 0.068 BTU/hr ft °F
Do = 0.0625 ft
Nu = 518.79

$$\boxed{ho = 564,\ 44\ \text{BTU/hr Pie}^2\ °F}$$

With all the necessary data for the equation of the global heat transfer coefficient are substituted to find its value, it is worth mentioning that the coefficient that will be obtained next will have to be related to the assumed one and recalculate the area again to observe the variation, the area obtained from the second calculation is the area that will be taken into account for the design of the new E-319 pentane cooler.

$$U = \dfrac{1}{\left(\dfrac{1}{hi} \times \dfrac{Ao}{Ai}\right) + \left(Rfi \times \dfrac{Ao}{Ai}\right) + Rw + \dfrac{1}{ho} + Rfo}$$

Data
hi = 195.44 BTU/hr ft °F^2
Ao = **As** = 461.37 Feet^2
Ai = 360, 03 Feet^2
Rw = 1.49 x 10^{-4}
Rfi =0.0015 hr °F foot /BTU^2
Rfo =0.0015 hr °F ft /BTU^2
ho = 564, 44 BTU/hr Foot °F^2

$$\boxed{U = 83,33\ \text{BTU/hr pie}^2\ °F}$$

When the difference between the two global heat transfer coefficients is observed, the area is recalculated again to obtain the real one and at the end of the chapter the data obtained will be compared with those of the simulation of the system.

$$\boxed{As = \dfrac{Q}{U \times \Delta T_{ml} \times F}}$$

Data
Q = 1,991,206.66 BTU/hr

$U = 83.33$ BTU/hr ft °F^2

$\Delta T_{ml} = 108{,}44$ °F

$F = 0{,}996$

Once the actual heat transfer area is known, the number of tubes is recalculated.

$$N^\circ \text{ de tubos} = \frac{As}{N^\circ \text{ de pasos} \times \pi D_o \times L}$$

Data
As = 221.22 Feet^2
Do = 0.0625 ft
L = 10 ft
No. of steps = 2

$$As = 221{,}22 \text{ Pies}^2$$

$$N^\circ \text{ de tubos} = 56{,}29 \sim 56 \text{ tubos}$$

The number of tubes obtained is 56 tubes of nominal diameter %", BWG 14 of carbon steel with a space between tubes or pitch of 1". The PDVSA standard allows a pipe spacing of 1.25 times multiplied by the external diameter, although exceptions can be made and take a pipe spacing of 1". Taking into account that 1.25 times the diameter of the % inch tube is 0.94" approximately, the exception can be made and round it to 1", since it is also a small exchanger.

At the same time, it should be noted that there are 56 two-pass tubes, i.e. 56 tubes going and 56 returning through the heat exchanger for a total of 112 tubes. A 20 to 25% overdesign is suggested, taking into consideration that these are small diameter tubes which can become clogged or erode over time during operation, as well as the variation of mass flows, temperature differentials, and/or unexpected pressure differentials that are outside the operational limits. From the number of tubes the casing diameter can be determined graphically.

Calculation of the diameter of the chiller casing.

The type of exchanger selected is a type exchanger (BKU).

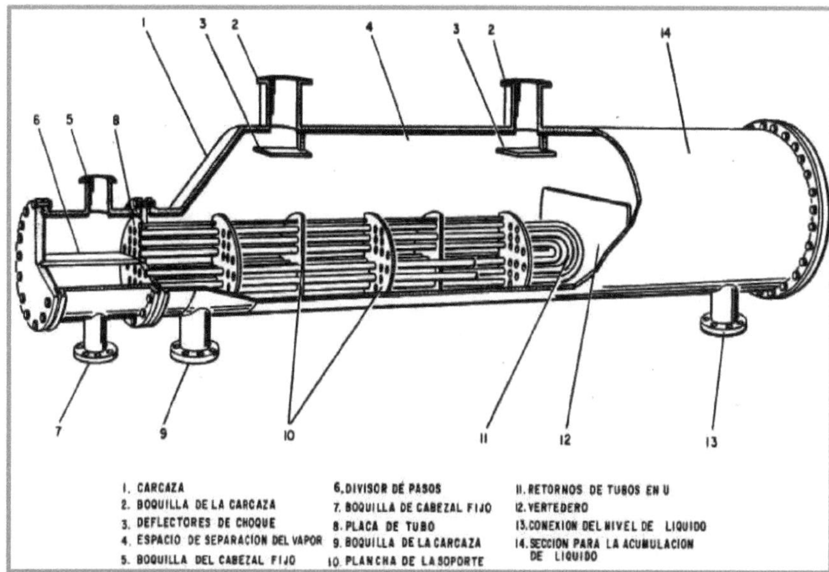

Figure 43. **BKU type exchanger with weir.** Source: PDVSA Standard (1995)

According to PDVSA MDP-05-E-02 Rev. 0, the coolers are normally configured as a kettle type exchanger but without weir (BKU). According to the nomenclature of the TEMA standard, it is an exchanger with a removable channel header with integral cover (Type B). The header is used with fixed tube plate, U-tube and removable bundle exchanger designs. This type of header is normally used only when the tube side fouling factor is less than 0.00035 m2°C/W (0.0020 hpie°F/BTU).

The housing is of the kettle type which is large enough and provides adequate space for the separation of the liquid and 126 The rear header is of the U-type, which indicates the construction of the tube bundle with U-tubes. Additionally, it should be noted that this type of heat exchanger is the most commonly used in the large fractionation plant in operations involving vaporization.

According to the following figure, the diameter of the plates supporting the tube bundle (support plates Fig. 41) should be determined and then a distance of 1" inch (25.4 mm) is added to the height of the previous head.

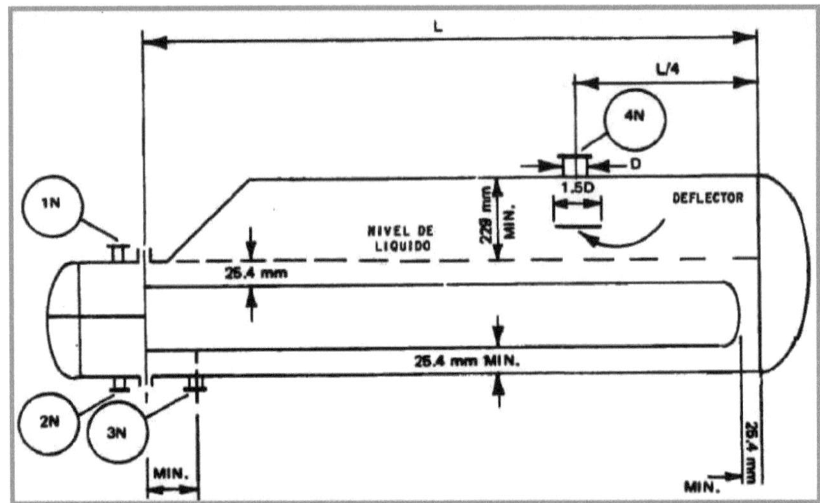

Figure 44. **Sizing of a BKU type exchanger.** Source: PDVSA Standard (1995)

The diameter of the support plates can be obtained from the following figure, the lower horizontal line represents the tube count, while the left vertical line symbolizes the diameter of the casing knowing the pitch and diameter of the tubes and they are represented by the descending curves. For the design of the new pentane cooler these data are known, in the case of the tubes the number of tubes is 112 up to, the space between them is 1" and their nominal diameter is ¾".

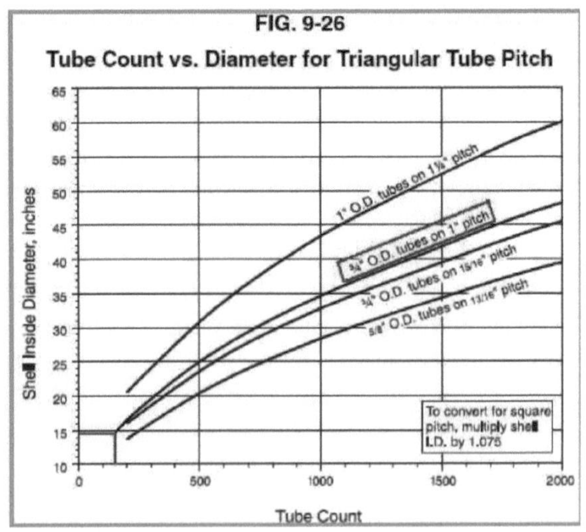

Figure 45. **Heat exchanger shell diameters.** Source: GPSA (2004)

Intercepting the number of tubes with the curve representing the characteristics of the tube bundle, it is observed that the approximate diameter is 14.5" ~ 15" inches added to 1" of separation between the bundle and the head height and 1" from the tube bundle to the bottom of the chiller is 17".

After obtaining this diameter, Figure 43 shows that a height of 229 mm (9") must be left as minimum space from the head height to the upper limit of the casing according to PDVSA MDP- 05-E-02 Rev. 0 standard, therefore 30% more than the minimum height is taken as a safety factor:

229 mm x 30% = 297,7 mm = 11,72" ~ 12" + 17" = 29"

Casing diameter = 29"

Calculation of chiller nozzle diameters

This procedure is based on the fluid velocity in the inlet lines to the cooler and the flow rate it handles, according to PDVSA L-TP 1.5 "Hydraulic Calculation of Pipelines" the general recommendation is an approximate velocity of between 5 and 15 ft/sec in lines that transport liquids, therefore this is the approximate velocity at which both the refrigerant (propane) and the product (pentane) will enter the new E-319 pentane cooler.

It should be noted that both velocities, the pentane and propane inlet velocities to E-319 are within the PDVSA standard, since both are around 5 ft/sec. First, the diameter of the inlet nozzle to E-319 of the propane refrigerant is calculated.

Calculation of coolant inlet nozzle diameter (casing diameter)

$$V_{max} = \frac{S_t}{S_t - Do} \times Vm$$

By clearing Vm

$$Vm = \frac{V_{max}(S_t - Do)}{S_t}$$

Data

St = 0.083 ft
Do = 0.0625 ft
Vmax = 5 ft/sec

$$Vm = 1{,}23 \text{ Pie / seg}$$

With the average velocity, the cross-sectional area of the **At** nozzle is found by clearing it from the following expression:

$$Vm = \frac{\dot{v}}{At} \qquad (12)$$

$$At = \frac{\dot{v}}{Vm}$$

Data

v = 1468 BPD = 0.0954 foot /sec^3
Vm = 1.23 ft / sec

$$At = 0{,}0776 \ pie^2$$

Now with the cross-sectional area of the nozzle, the diameter is finally found by clearing it from the following expression:

$$At = \frac{\pi}{4} \times D_i^2 \qquad (13)$$

$$D_i = \sqrt{\frac{At}{\pi/4}}$$

Data

At = 0.0776 ft^2
π = 3,14

$$D_{boquilla} = 0{,}315 \ pies = 3{,}78 \ pulg \sim 4 \ pulg$$

Calculation of pentane inlet nozzle diameter (tube inlet diameter)

To find the inlet diameter of the pentane to the tubes, the following equation is applied, from which the area **At** is cleared:

$$Q = V \times At$$

Data

Q = **flow rate** = 76762 Lb/hr = 8360 BPD = 0.5433 Foot^3 / sec
V = **fluid velocity** = 5 ft/sec

$$At = 0.109 \ ft^2$$

To find the diameter, the following expression is used to define the casing inlet diameter

$$D = 0{,}37 \ pies = 4{,}48 \ pulg \sim 5"$$

$$D_i = \sqrt{\dfrac{At}{\pi/4}}$$

Data

$A = 0.109$ ft^2

$\pi = 3,14$

Calculation and selection of cut, spacing and number of baffles

Cutting of the deflector:

According to the type of heat exchanger, the baffles must be oriented perpendicular to the tube bundle axis with a vertical cut as shown in Figure 41.

According to PDVSA standard MDP-05-E-02 Rev. 0, the baffle cuts are usually at 25% of the casing diameter, the support plates must be cut to allow the coolant flow through the casing, for this reason the cut will be made based on the recommended PDVSA standard.

25% x 17" = 4.25" = **Horizontal length of deflector**

Space between deflectors

The minimum baffle spacing per PDVSA standard required to maintain good flow distribution is 20% of the casing inside diameter but not less than 50 mm (2 in.).

Too small a baffle spacing forces the fluid in the casing to deflect, thus producing a decrease in the heat transfer coefficient. While the maximum baffle spacing should not exceed the casing diameter for this reason the spacing it will take will be 50% of the casing diameter.

50% x 17" = 8.5" = **Space between baffles**

The number of deflectors is deduced from the above result since the length of the tubes is 10 feet and the deflectors will be distributed along the length of the tube bundle, therefore the number of deflectors is:

Pipe length = 10 feet = 120 inches/8.5 inches = 14

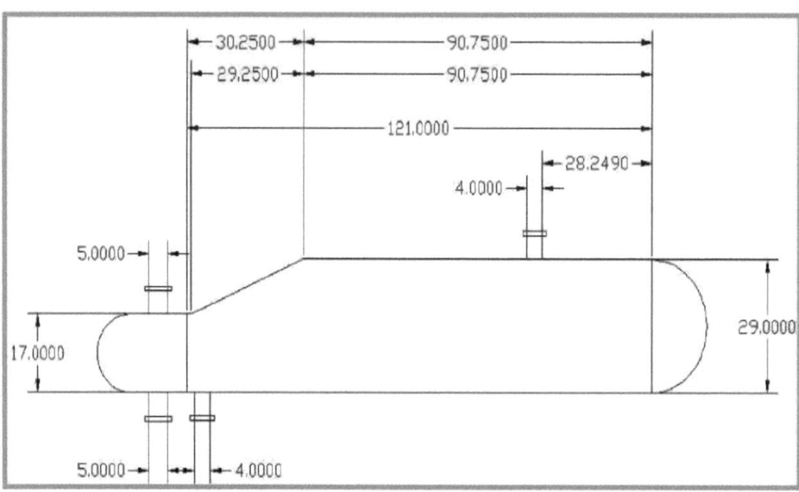

Figure **46 Dimensions of the new cooler E-319.** Image obtained using the AutoCAD 2007 design program. Source: Mendoza and Barreto (2011)

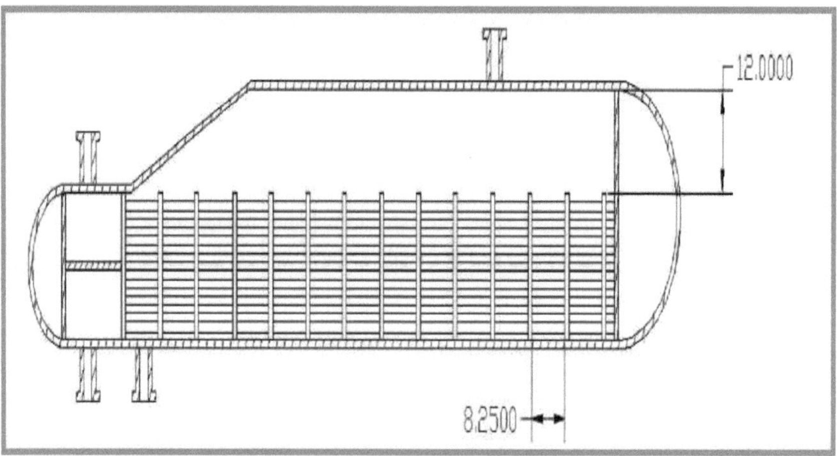

Figure **47 Longitudinal section of the new cooler E-319.** Image obtained using the AutoCAD 2007 design program. Source: Mendoza and Barreto (2011)

New cooling system scheme of the Bajo Grande LPG plant with the new E-319 pentane chiller.

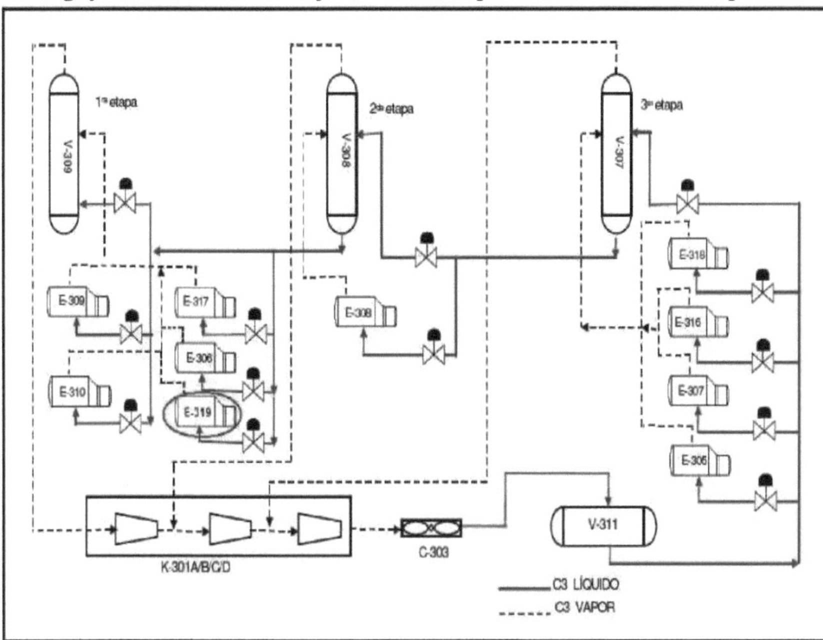

Figure 48. **New cooling system scheme of the Bajo Grande LPG plant with the new E-319 pentane chiller.** Source: Mendoza and Barreto (2011)

After the calculations and analysis of the site where the new E-319 pentane chiller will be located, we proceed to describe the cooling system of the plant under large with the new E-319 pentane chiller.

The propane leaving the compressor discharge K-301A/B/C/D is condensed in the air coolers C-303 and then collected in the accumulator drum V-311; from there it is fed to the casings of the following coolers, located at the first temperature level: E-305 (propane), E-318 (n-butane), E-316 (i-Butane) and E-307 (gasoline). In each of the coolers, the propane vaporizes by receiving the heat from the products being cooled. The vapors resulting from the evaporation of propane from each cooler are collected in a common line and carried in the V-307 drum.

From the accumulator V-311, there is another liquid propane line which expands before reaching the economizer V-307. Its function is to guarantee the vapor flow to the third compression stage. From V-307 liquid propane flows to cooler E-308 (n-Butane cooler), located at the second temperature level. The vapor leaving the cooler goes to the economizer V-308. As in the first temperature level, liquid propane is extracted from V-307, which expands before reaching economizer V-308 to ensure vapor flow to the second compression stage.

The propane vapors generated in the E-308 cooler go to the V-308 economizer where they are mixed and constitute the suction of the second compression stage. The liquid propane from economizer V-308 feeds the third temperature level of the refrigeration system, consisting of the coolers: E-306 (propane), E-317 (i-Butane), E-309/E-310 (vapors from tank S-501), E-319 (New pentane cooler). The inlet conditions of the propane refrigerant in the new pentane chiller are -14.8

°F and 17.71 Psig and exits at a temperature of -9 °F and 17.21 Psig in vapor phase, the pentane in the heat exchange is cooled from 120.5 °F to 75 °F.

The vapors produced in the aforementioned coolers go to the suction drum of the first stage of the V-309 compressor, thus closing the compression cycle and continuing to circulate through the process. To ensure vapor flow to the first compression stage, a line of liquid propane is drawn off and expanded before reaching the V-309.

Verification of results obtained with Hysys 7.2 simulator

Note 5

The compositions, properties, conditions and other characteristics of the flow streams specified for the simulation are in Annex 1, since it is a report obtained in Hysys automatically and is fully specified for a better appreciation.

Note 6

Hysys classifies input and output data as follows:
•   With Blue color classifies the data that are specified by the designer, these data can be modified depending on the characteristics of the process and/or by convenience of the designer.

•   In the case of the heat exchanger, Hysysys has as default data suggested by the TEMA standard (e.g. shell diameter, number of tubes, tube spacing, among others), these are basic design data of heat exchangers and the most used at industrial level, these data can also be modified both by internal calculations of the simulator and by the designer.

•   Finally, Hysys classifies with Black color all those data that cannot be modified, that is, data that are obtained directly from Hysys database equations when some parameters are specified, as well as from thermodynamic packages under which the simulator operates.

The data entered in the simulator were the minimum required by Hysys to perform the calculations, among these are the temperatures, pressures of each fluid, in the case of the mass flow only the mass flow of the fluid was specified.

pentane since the propane one was not known and was left to be calculated by the simulator to rectify it at the end. The number of tubes, tube spacing, tube bundle arrangement, pentane and propane composition, film coefficients of both fluids, fouling factor were also specified. The rest of the data were generated by the simulator.

After entering all the data in the simulator, the following results were obtained

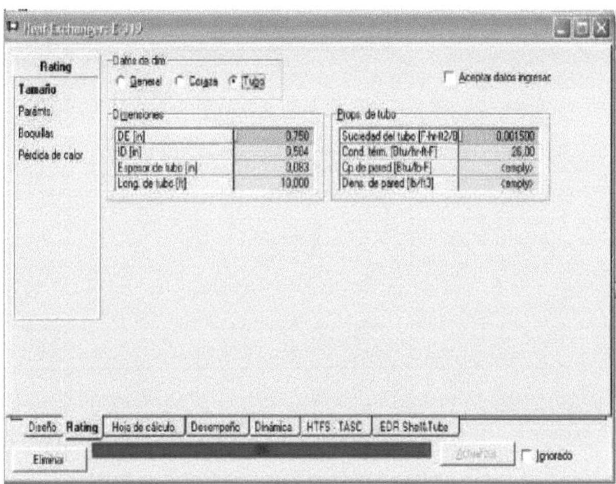

Figure 49. **Specification of pipe dimensions (see note 6).** Source: data obtained using the Hysys 7.2 simulator. Mendoza (2011)

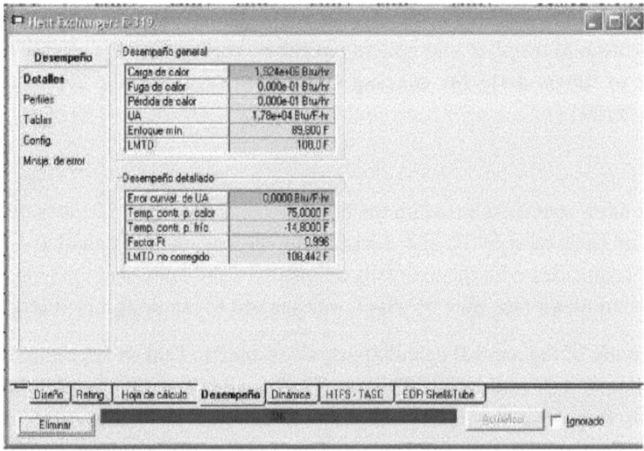

Figure 50. **Heat load, LMTD and Correction Factor "F" generated by the simulator (see note 6).** Source: data obtained in Hysys 7.2. Mendoza (2011)

It can be seen that the heat load is similar to that obtained in the calculations approximately 1.99 MMBTU/ hr the heat flux obtained in Hysys is approximately 1.92 MMBTU/hr. It can also be seen the similarity of the log mean temperature difference (LMTD) generated by the simulator 108.442 °F compared to the calculations is the same, finally the correction factor F likewise agrees with those obtained in the calculations 0.996.

The next step is the physical part of the heat exchanger, which is directly related to what has been previously verified.

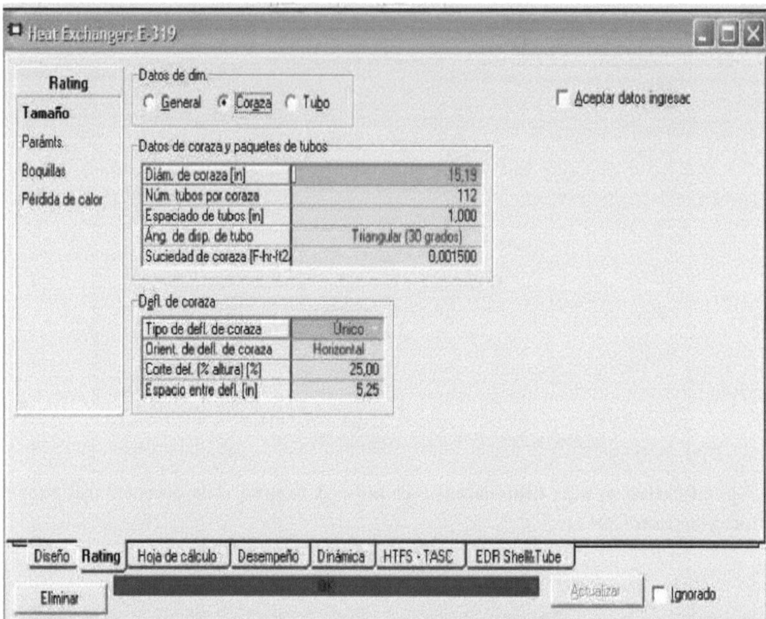

Figure 51. **Specification of number and spacing of tubes, casing diameter, internal fouling factor and arrangement of tubes, deflector spacing and cutout (see note 6).** Source: Data obtained in Hysys 7.2. Mendoza (2011)

The number of tubes specified based on the manual calculations is 112 tubes of %" with a center to center spacing of each tube of 1", the diameter of the casing was calculated by the simulator approximately 15" (coincides with the manually calculated diameter), also the type of predetermined arrangement of the simulator (see note 6) which was applied to the design is observed.

From the beginning of the manual calculations, we wanted to find an approximation of the global heat transfer coefficient "U", which would allow us to have a notion of what the transfer area would really be, in order to then develop a series of manual calculations, which must be verified.

For this reason, the value of the global heat transfer coefficient "U" generated by the simulator is presented below, when all the manual data are specified in the simulator (see note 6).

The value of this coefficient is taken as a reference because it depends on all the calculations made (hence its name), i.e., if the results show errors, the simulator will give a very different result from the manual calculations.

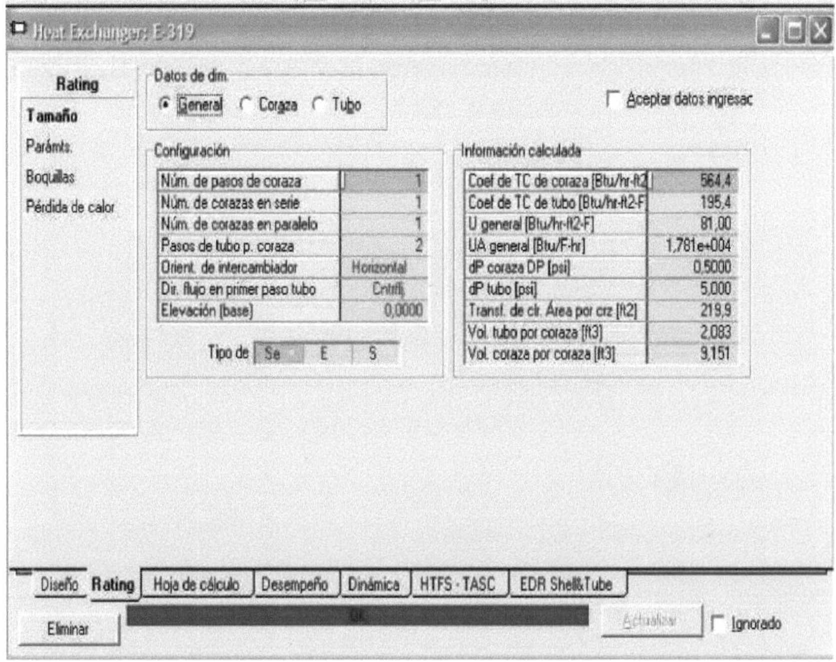

Figure 52. **Overall heat transfer coefficient (see note 6).** Source: data obtained in Hysys 7.2. Mendoza (2011)

When entering the data obtained manually in the simulator, a value of the global heat transfer coefficient of 81 BTU/hr ft^2 °F was obtained, being 83.33 BTU/hr ft^2 °F the U calculated manually (see calculations of the sizing of the new pentane cooler), if we take into account all the calculations made, interpolations made, approximations in graphs, among other aspects, this difference is considered negligible, although if another iteration was made to obtain the transfer coefficient, the value calculated by the simulator could be reached.

The mass flow of the refrigerant required to cool the pentane to the desired temperature is then compared with the manual result. The mass flow was calculated directly by the simulator and is considered similar to that obtained in the manual calculations 10860 Lb/hr. Hysys applies the equation of the first law of thermodynamics for a heat exchanger by the difference of the input and output enthalpies of the fluid calculates the required volume of refrigerant (See note 6).

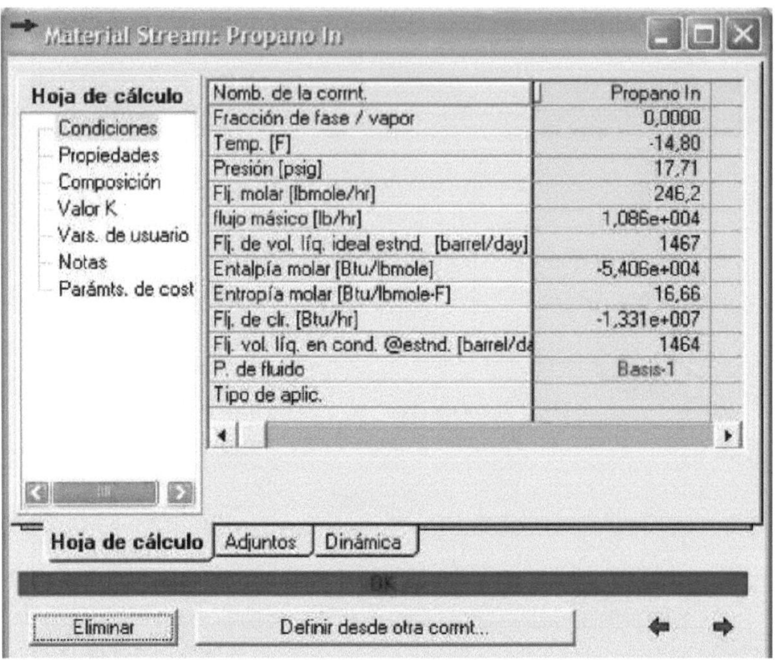

Nomb. de la cormt.	Propano In
Fracción de fase / vapor	0,0000
Temp. [F]	-14,80
Presión [psig]	17,71
Flj. molar [lbmole/hr]	246,2
flujo másico [lb/hr]	1,086e+004
Flj. de vol. líq. ideal estnd. [barrel/day]	1467
Entalpía molar [Btu/lbmole]	-5,406e+004
Entropía molar [Btu/lbmole-F]	16,66
Flj. de clr. [Btu/hr]	-1,331e+007
Flj. vol. líq. en cond. @estnd. [barrel/da	1464
P. de fluido	Basis-1
Tipo de aplic.	

Figure 53. **Refrigerant mass flow (see note 6).** Source: data obtained in Hysys 7.2. Mendoza (2011)

The verification of the results is concluded (see note 5) and the length and characteristics of the pentane and propane transmission lines are calculated.

Calculation of the length of the supply and outlet piping for the propane refrigerant of the E-319 chiller.

For the design of the refrigerant supply piping it is taken into account that the inlet diameter of the nozzle of the new cooler is 4", for this reason the diameter of the inlet and outlet piping of the cooler will have these dimensions. It will be connected to the propane supply line coming from V-308. And based on this the approximate length of the line to the new pentane cooler is calculated.

Using google map software, the distances are as follows:

A = 10 m from the propane refrigerant accumulator to the refrigerant inlet of the E-319 chiller.

B = 9 mts. from the coolant outlet of the cooler E-319 to connect with the suction drum line V-309.

Figure 54. **Calculation of the approximate length of the E-319 chiller refrigerant propane supply and discharge piping** Source: data obtained using Google Map (2011). Mendoza (2011) **Calculation of pentane supply and outlet piping**

The type of pipe used for the calculations will be carbon steel, Schedule 40 and its diameter will be the same as that of the pentane outlet nozzles of the E-319 cooler, 5 inches. When calculating the distance from the depentanizer unit to the new E-319 "C" pentane cooler using google map the distances are as follows:

C = 110 mts. Pentane supply line from the depentanizer.

D = 160 mts. Exit line of the E-319 cooler pentane.

Figure 55. **Calculation of the approximate length of the pentane supply and discharge piping in the E-319 cooler.** Source: data obtained using Google Map (2011). Mendoza (2011)

In this case it is convenient to simulate the line due to its considerable length. PipePhase 9.1 software will be used for this procedure. For the simulation an ambient temperature of 90 °F is considered (and it is assumed that the inlet and outlet pipe section of the cooler will be exposed to the environment (Pipe in Air).

The nomenclature used is as follows: **C5-S** is the exit point of the pentane to the tank, **G000** is the arrival point of the pentane to the tank, the line between them represents the pipe section between them and its identification is **P001**.

Similarly, **C5-E** represents the exit point from the depentanizer to E-319; and **D000** is the arrival point at the cooler; the line between both is the pipe section between the depentanizer and the cooler E-319 and according to the default configuration of the simulator its identification is **P003**.

Prior to obtaining the results, all known data were specified for the simulation of the pipe, the approximate distance is 110 meters, the nominal diameter is 5 inches, Schedule 40, the standard temperature of the area is 90 °F, it is assumed that the pipe will be exposed to the environment (pipe in air) as shown in Figure 53.

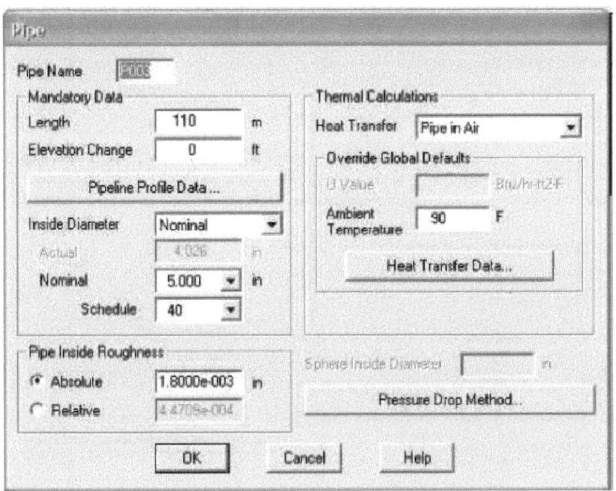

Figure 56. **Specification of the characteristics of the pipe coming from the depentanizer unit to E-319.** Source: data obtained using the PipePhase 9.1 simulator. Mendoza (2011).

The outlet conditions from the depentanizer unit is 120.5°F and 68.5 Psig and the specified mass flow rate is 76762 Lb/hr. As shown in Figure 54.

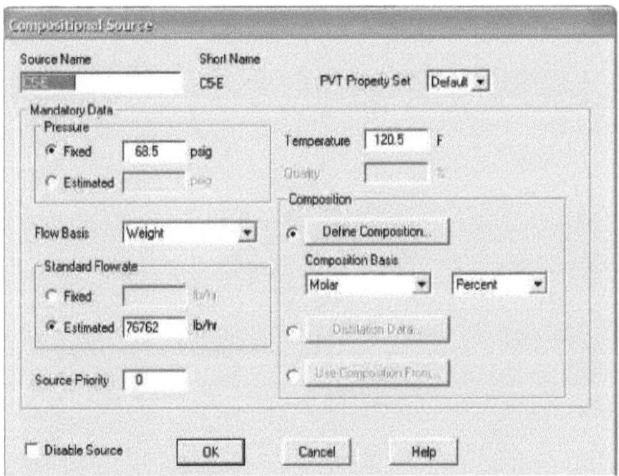

Figure 57. **Specification of the conditions and properties of the pentane coming from the depentanizer unit.** Source: PipePhase 9.1. Mendoza (2011).

Next, the data corresponding to the pipe line that will carry the pentane to the storage tank were specified, its dimensions are 160 meters long, 5 inches, Schedule 40 carbon steel, likewise the ambient temperature is 90 °F and it will be exposed to the environment.

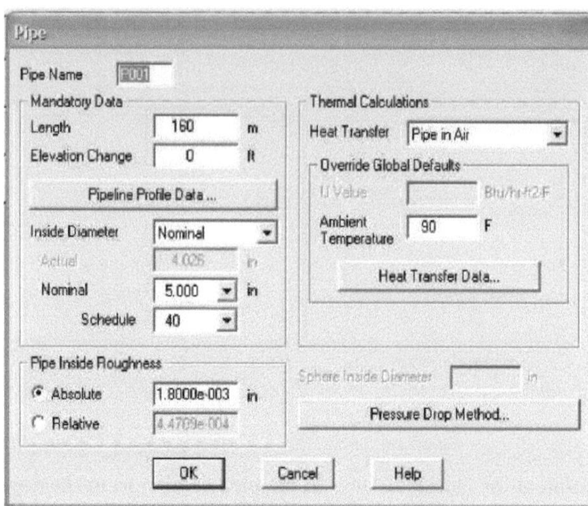

Figure 58. **Specification of pentane outlet piping characteristics of E-319.** Source: PipePhase 9.1. Mendoza (2011).

Finally, the pentane outlet conditions of the E-319 cooler at a pressure of 63.5 Psig, 75 °F temperature and 76762 Lb/hr were specified.

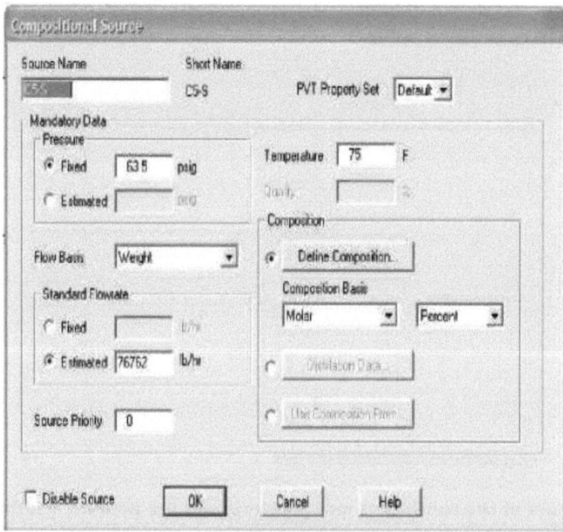

Figure 59. **Specification of pentane conditions and properties at the exit of E-319.** Source: PipePhase 9.1. Mendoza (2011).

The results obtained in the simulation are shown below:

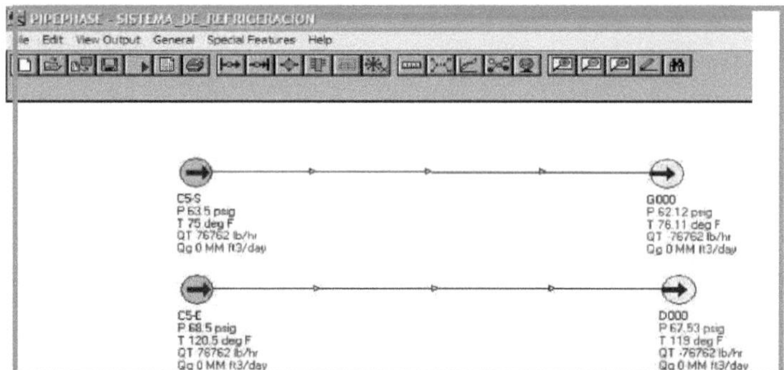

Figure 60. **Simulation of pentane supply and discharge lines in E-319.** Source: PipePhase 9.1. Source: data obtained through the PipePhase 9.1 simulator. Mendoza (2011).

In the following graph the pressure drop in the pentane transfer pipe to the cooler is 1 Psig, i.e. 9.09×10^{-3} per meter of pipe, it can be considered negligible without influence to make any modification to the conditions of entry to the cooler considering that these conditions are ideal but not real, in this sense the procedure and the results are estimated for this reason is always considered the overdesign.

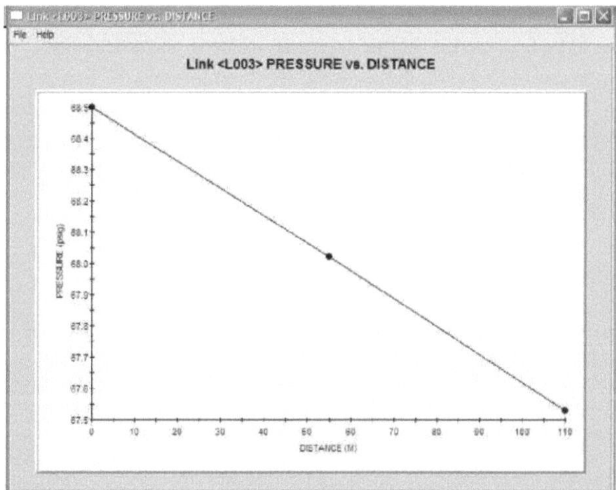

Figure 61. **Results obtained in the simulation: "Pressure Vs distance graph of the pipe coming from the depentanizing unit.** Source: data obtained using the PipePhase 9.1 simulator. Mendoza (2011).

The following figure is the graph of temperature Vs. distance, it shows the blue line at the bottom of the graph, this symbolizes the ambient temperature line 90 °F, while at the top is the black line and this represents the temperature of pentane at 120.5 °F, this line has a downward trend because the pentane gives up heat to the atmosphere at a rate of 0.01 °F per meter of pipe.

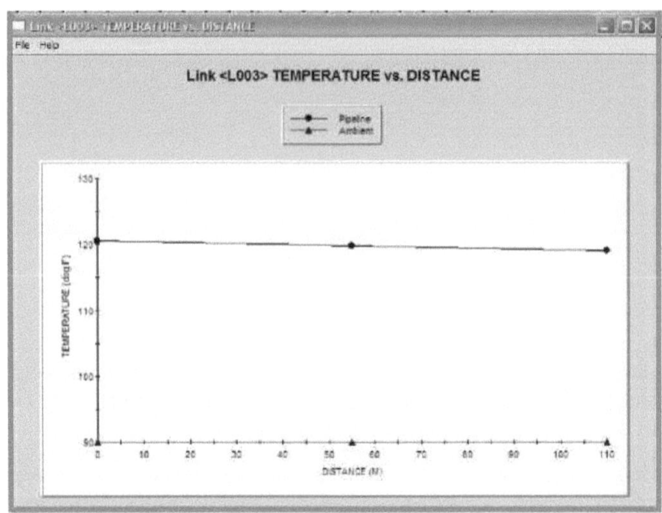

Figure **62 Results obtained in the simulation: "Temperature Vs distance graph of the pipe coming from the depentanizing unit.** Source: data obtained using the PipePhase 9.1 simulator. Mendoza (2011).

The first variable to be analyzed is the fluid velocity since in the calculations for the sizing of the cooler the velocity was around 3 ft/sec, in this sense when simulating the line at the given conditions this velocity is very close to the calculated velocity, the result is 4 approximately at the inlet and outlet of the cooler. The pressure drop from the inlet pipe to the cooler was 1 psig (P003), while from the cooler outlet to the pentane storage tank (P001) there is a pressure drop of 1.4 psig this can be better observed in Figures 58 and 62.

```
                            VELOCITY SUMMARY

                                              PRESSURE
          DEVICE DEVICE  MIXTURE VELOCITY  CRITICAL    GRADIENT       PRESSURE
    LINK  NAME   TYPE    (INLET/OUTLET)    VELOCITY    (INLET/OUTLET) DROP
                        (FPS)             (FPS)       (PSIFT)        (PSIG)

    ---- ------ ------  --------------- --------- --------------- --------
    L000 P001   PIPE    3.96    3.96     0.00 -2.6E-3 -2.6E-3      -1.4
    L003 P003   PIPE    4.14    4.13     0.00 -2.7E-3 -2.7E-3      -1.0
  ■ VERSION 9.1                                    R
    SIMULATION SCIENCES, INC.        PIPEPHASE                     PAGE  12
    PROJECT                          OUTPUT
    PROBLEM                          NETWORK REPORT                10/23/11
```

Figure **63 Results obtained in the simulation: "Fluid velocity and pressure drop inside the pipeline coming from the depentanizing unit V-8331.** Source: data obtained using the PipePhase 9.1 simulator. Mendoza (2011)

In the pressure Vs distance graph it is observed that the pressure drop is directly proportional to the length of the pipe, i.e., the longer the pipe the greater the pressure drop, the pressure gradient is approximately 8.75 x 10^{-3} Psig per meter of pipe for a total of 1.4 Psig. Operationally this pressure drop can be considered as negligible if we take into account the length of the pipe. In addition, the pdvsa standard recommends a maximum pressure drop of 4 Psig per 100 feet of pipe as shown below:

	PROCEDIMIENTO DE INGENEIRÍA		PDVSA L-TP 1.5	
PDVSA	CÁLCULO HIDRÁULICO DE TUBERÍAS		REVISION 0	FECHA JUL.94
			Página 45	

Menú Principal	Índice manual	Índice volumen	Índice norma

TABLA 3

VELOCIDAD RECOMENDADA Y ΔP MAXIMA PARA LIQUIDOS EN TUBERIAS DE ACERO AL CARBONO

TUBERIAS PARA LIQUIDOS EN SERVICIO DE PROCESO Y EQUIPOS

Tipo de Servicio	Velocidad Pie/seg.	ΔP Máximo Lppc/100 Pies
1. Recomendación General	6 – 15	4
2. Fluio Laminar	4 – 5	

Figure **64 PDVSA standard "hydraulic calculation of pipelines".** Source: PDVSA (1994)

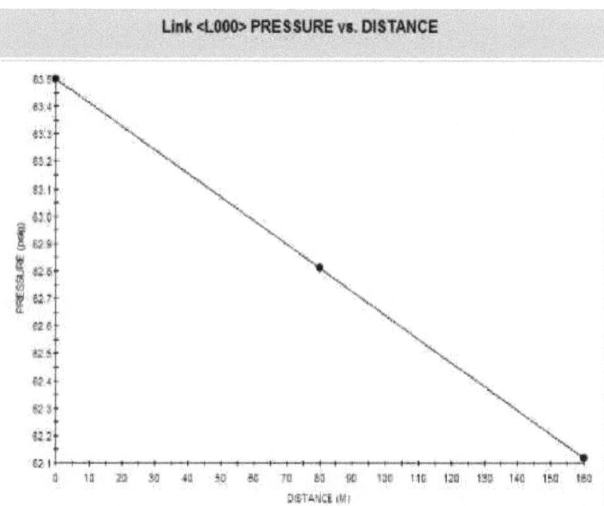

Figure **65 Results obtained in the simulation: "Pressure Vs distance graph of the outlet pipe from the E-319 cooler to the pentane storage tank.** Source: data obtained using the PipePhase 9.1 simulator. Mendoza (2011)

The following graph shows the blue line at the top representing the 90°F ambient temperature line, at the bottom at 75°F is the pentane temperature line (black line).

It can be noticed that the black line has an upward trend that tends to seek the blue line, this means that as the fluid moves along the pipe, the temperature tries to equilibrate with the ambient temperature as it provides heat to the pipe, causing an increase in the pentane temperature from 75°F to approximately 76.11°F.

Operationally, this tank arrival temperature may cause thermal shocks inside the tank, but it should be noted that the pentane also loses heat before entering the cooler.

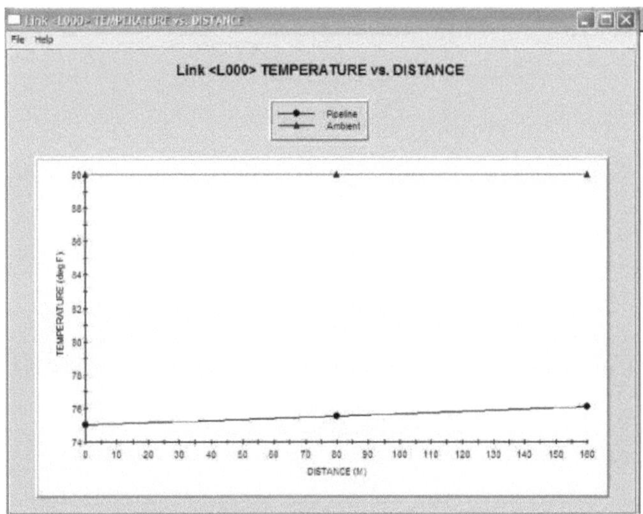

Figure 66. **Results obtained in the simulation: "Temperature Vs Distance graph of the outlet pipe from the E-319 cooler to the pentane storage tank.** Source: data obtained using the PipePhase 9.1 simulator. Mendoza (2011)

CONCLUSIONS

Based on the results obtained, the following conclusions have been reached:

- The existing refrigeration system at the large fractionation plant meets the thermal and service requirements for the installation of the new E-319 pentane chiller.

- The operational parameters of the new E-319 pentane cooler were defined as follows:

E-319 cooler operational parameters					
	Pentane			**Propane**	
Conditions		**Entrance**	**Output**	**Entrance**	**Output**
Pressure (Psig)		68,5	*63,5*	17,71	17,21
Temperature (^{0} F)		120,5	*75*	-14,8	-9
Mass Flow (Ib/hr)		*76762*	76762	10860	10860

The new pentane cooling system will consist of the following elements:

- One BKU type heat exchanger, two (2) pass through tubes and one (1) pass through shell, the tube material shall be carbon steel with length of 10 feet, BWG 14 gauge corresponding to nominal diameter %". The tubular arrangement shall be 30° triangular. The header type is removable channel with integral cover (Type B). The rear head is U-type, the same indicates the U-shaped tube bundle construction. While the product inlet and outlet nozzles are 5" and 4" for coolant.

- . The product transfer piping for inlet and outlet of pentane is carbon steel, Schedule 40 with a nominal diameter of 5" and 110 meters long for the inlet and 160 meters long for the outlet of the product to the tanks. The associated lines for the refrigerant supply are 10 mts. for the supply and 9 mts. for the refrigerant outlet to the V-309.

RECOMMENDATIONS

For the proper operation of the proposed cooling system at the Bajo Grande fractionation plant, the following is suggested:

• Coat the product transfer piping from the depentanizer unit to the new E-319 pentane cooler, and from there to the storage tanks to avoid off-spec product.

• It is recommended that a suitable pumping system be selected to handle the flow of pentane to the tanks.

• Conduct a study to know the influence on the existing units in the refrigeration system resulting from the installation of E-319, especially on the refrigeration level selected for the location of the new pentane chiller.

• It is suggested to add the minimum mass flow of 10860 lb/hr required to the cooling system for pentane cooling in order not to produce variations in the mass flows of the existing chillers at the selected cooling level for the E-319 location.

• A technical-economic study should be carried out to select the most feasible route for the pentane transmission lines to and from the E-319 cooler, to avoid unfavorable operational conditions that influence the stability of the process.

BIBLIOGRAPHY .

Kern, Donald Q (1999). Heat transfer processes. Thirty-first reprint. Editorial McGraw Hill. Mexico D.F.

Cengel, Yunus (2007). Heat and Mass Transfer. Third Edition. Editorial McGraw Hill/ Interamericana de Editores, S.A. de C.V. Mexico City.

Cao, Eduardo (2004). Heat Transfer in Process Engineering. First Edition. Buenos Aires, Argentina.

Cengel, Yunus and Boles, Michael A (2009). Thermodynamics. Sixth Edition. Editorial McGraw Hill/ Interamericana de Editores, S.A. de C.V. Mexico City.

Pérez, Ramiro and Martínez Macías (1994). Characteristics and Behavior of hydrocarbons. Consulting Engineers. Maracaibo Venezuela.

GPSA Gas Processors Suppliers Association (2004). Engineering Databook 12th Edition. FPS Version, Volumes I & II. Tulsa, Oklahoma, United State of America.

TEMA Tubular Exchanger Manufacturers Association (1999). 8th Edition. 25 North Broadway Tarrytown, New York.

Abd Hamid, M.K. (2007) *"HYSYS: An Introduction to Chemical Engineering Simulation"*. Malaysia University Technology.

Hernández, R et al (2000). Research Methodology. Second Edition. Editorial Mc Graw Hill. México.

Hurtado, J (2004). Holistic Research Methodology. Third Edition. Editorial SYPAL. Caracas, Venezuela.

PDVSA (1995). Process Design Manual MDP-01-DP-01. Design Temperature and Pressure. Venezuela.

PDVSA (1995). Process Design Manual MDP-02-FF-02. Flow of Fluids. Basic Principles. Venezuela.

PDVSA (1995). Process Design Manual MDP-05-E-01.... Heat Transfer. Heat Exchangers. Basic Principles. Venezuela

PDVSA (1995). Process Design Manual MDP-05-E-02 Heat Transfer. Shell and Tube Heat Exchangers. Venezuela.

PDVSA (1994). Design Engineering Manual L-TP 1.5. Volume 13-III Hydraulic Calculation of Pipelines. Venezuela

PDVSA (1994). Design Engineering Manual 90616.1.024. Volume 13-III Process Piping Sizing. Venezuela

PDVSA (1994). Design Engineering Manual H-221. Volume 13-I Piping Materials. Venezuela

PDVSA Gas (2010). Western Cryogenic Complex Project (CCO). Evaluation Study of the Pentane Management of the CCO in Bajo Grande Doc. No. CCO-PIPC-03-PR-EST-10-0003 Rev. 0. Venezuela.

PDVSA Gas (2010). Western Cryogenic Complex Project (CCO). Evaluation Study of the Existing Refrigeration System of the Bajo Grande LPG Plant, to Consider the Installation of a New Pentane Product Chiller. Doc. No. CCO-PIPC-03-PR- EST-10-0005 Rev. 0. Venezuela.

PDVSA Gas (2007). Western Cryogenic Complex Project (CCO). Description of the Block VII Process. Doc. No. 1150-01-40-P01-TEC- 003 Rev. 0. Venezuela.

PDVSA Gas (2007). Western Cryogenic Complex Project (CCO). Mass and Energy Balance Fractionation Unit Doc. No. 8922Y-400-CN-0001-01. Rev. 0. Venezuela.

PDVSA Gas (2007). Operations Manual of the Bajo Grande LPG Fractionation Plant. Doc. No. POPE 006. Rev. 1. Venezuela.

PDVSA Gas (2007). Western Cryogenic Complex Project (CCO). Process Flow Diagram of the V-8331 Depentanizing Unit of the Bajo Grande Fractionation Plant. Doc. No. 1150-01-70- P23-101 Rev.0. Venezuela.

Printed by Books on Demand GmbH, Norderstedt / Germany